Johannes Rau

Tropical intersection theory and gravitational descendants

Johannes Rau

Tropical intersection theory and gravitational descendants

Intersections of tropical cycles and applications to enumerative geometry

Südwestdeutscher Verlag für Hochschulschriften

Impressum/Imprint (nur für Deutschland/ only for Germany)
Bibliografische Information der Deutschen Nationalbibliothek: Die Deutsche Nationalbibliothek
verzeichnet diese Publikation in der Deutschen Nationalbibliografie; detaillierte bibliografische
Daten sind im Internet über http://dnb.d-nb.de abrufbar.
Alle in diesem Buch genannten Marken und Produktnamen unterliegen warenzeichen-, marken-
oder patentrechtlichem Schutz bzw. sind Warenzeichen oder eingetragene Warenzeichen der
jeweiligen Inhaber. Die Wiedergabe von Marken, Produktnamen, Gebrauchsnamen,
Handelsnamen, Warenbezeichnungen u.s.w. in diesem Werk berechtigt auch ohne besondere
Kennzeichnung nicht zu der Annahme, dass solche Namen im Sinne der Warenzeichen- und
Markenschutzgesetzgebung als frei zu betrachten wären und daher von jedermann benutzt
werden dürften.

Verlag: Südwestdeutscher Verlag für Hochschulschriften Aktiengesellschaft & Co. KG
Dudweiler Landstr. 99, 66123 Saarbrücken, Deutschland
Telefon +49 681 37 20 271-1, Telefax +49 681 37 20 271-0
Email: info@svh-verlag.de
Zugl.: Kaiserslautern, TU, Dissertation, 2009

Herstellung in Deutschland:
Schaltungsdienst Lange o.H.G., Berlin
Books on Demand GmbH, Norderstedt
Reha GmbH, Saarbrücken
Amazon Distribution GmbH, Leipzig
ISBN: 978-3-8381-1428-6

Imprint (only for USA, GB)
Bibliographic information published by the Deutsche Nationalbibliothek: The Deutsche
Nationalbibliothek lists this publication in the Deutsche Nationalbibliografie; detailed
bibliographic data are available in the Internet at http://dnb.d-nb.de.
Any brand names and product names mentioned in this book are subject to trademark, brand
or patent protection and are trademarks or registered trademarks of their respective holders.
The use of brand names, product names, common names, trade names, product descriptions
etc. even without a particular marking in this works is in no way to be construed to mean that
such names may be regarded as unrestricted in respect of trademark and brand protection
legislation and could thus be used by anyone.

Publisher: Südwestdeutscher Verlag für Hochschulschriften Aktiengesellschaft & Co. KG
Dudweiler Landstr. 99, 66123 Saarbrücken, Germany
Phone +49 681 37 20 271-1, Fax +49 681 37 20 271-0
Email: info@svh-verlag.de

Printed in the U.S.A.
Printed in the U.K. by (see last page)
ISBN: 978-3-8381-1428-6

Contents

Preface

Tropical geometry

Tropical geometry is a rather new field of mathematics whose general idea is similar to toric geometry: Geometric objects and properties are replaced by combinatorial ones. The connection between classical and tropical geometry is made by various "tropicalization" processes that transform (algebraic or symplectic) geometric objects into polyhedral sets satisfying a so-called balancing condition. The hope is that the polyhedral sets are easier to study and inherit enough properties from the original objects such that the tropical results can be transferred back and lead to new insights in classical geometry. One should mention that throughout this text we use the term "classical" as the opposite of "tropical" to refer to the usual algebraic (or symplectic) geometry — even though, for example, the "classical" theory of Gromov-Witten invariants is not older than 20 years and therefore nearly as modern as tropical geometry.

Tropical methods have proven to be successful in a number of different areas of mathematics, such as (real) enumerative geometry (cf. [Mi03], [IKS04]), symplectic geometry (cf. [Ab06]), number theory (cf. [Gu06]) on the one hand, and combinatorics (cf. [Jo08]), algebraic statistics and computational biology (cf. [PS03]) on the other hand. In this thesis, we follow the combinatorial approach to tropical geometry, i.e. we start with polyhedral complexes and piecewise affine functions and study their properties by means of convex and combinatorial geometry. However, the inspiration for the definitions and statements we prove is due to the classical world — algebraic intersection theory (cf. [Fu84]) in the first chapter, and Gromov-Witten theory in the second chapter.

Enumerative geometry

The most popular question of classical enumerative geometry is the following: How many complex plane rational curves of given degree d interpolate a generic configuration of points (the number of points is chosen such that we expect a finite number)? As long as two decades ago, this problem was solved only for few and small degrees. It was in 1994 (and inspired by predictions from string theory!) when a breakthrough was made by Kontsevich, who found a recursive formula for these numbers (cf. [KM94]) using intersection theory on the moduli space of stable maps. Based on this spectacular result, a rich theory of stable maps and Gromov-Witten invariants emerged. In particular, mathematicians (and physicists) got interested in "gravitational descendants" (the physical origin is obvious here), natural generalizations of Gromov-Witten invariants provided by imposing not only incidence conditions (such as the mentioned point conditions), but also "Psi-class" conditions. These "Psi-classes" are the first Chern classes of "canonical" line bundles of the moduli space (more precisely, pull backs of the cotangent bundle of the universal family along a section).

It was about ten years later, when Mikhalkin, based on a conjecture of Kontsevich, established a connection between the Gromov-Witten invariants (without Psi-classes) and the count of certain piecewise linear graphs in the plane — and thereby founded tropical geometry as a subject on its own (cf. [Mi03]). These results are based on deformations of the complex structure and other symplectic methods, and had implicit consequences such as the invariance of the tropical counts from the chosen (generic) point configuration. Later on, various attempts were made to understand these consequences without the symplectic methods, but inside tropical geometry (cf. [GM05]).

The results of this thesis

This thesis is devoted to two main topics (accordingly, there are two chapters):

1. We establish a *tropical intersection theory* with analogue notions and tools as its algebro-geometric counterpart. This includes tropical cycles, rational functions, intersection products of Cartier divisors and cycles, morphisms, their functors and the projection formula, rational equivalence. The most important features of this theory are the following:

    - It unifies and simplifies many of the existing results of tropical enumerative geometry (cf. [GM05] and [FM08]), which often contained involved ad-hoc computations.

    - It is indispensable to formulate and solve further tropical enumerative problems (cf. [KM06] and chapter 2).

    - It shows deep relations to the intersection theory of toric varieties and connected fields (cf. [FS94] and section 1.6).

    - The relationship between tropical and classical Gromov-Witten invariants found by Mikhalkin is made plausible from *inside* tropical geometry.

    - It is interesting on its own as a subfield of convex geometry.

2. We study tropical *gravitational descendants* (i.e. Gromov-Witten invariants with incidence and "Psi-class" factors) and show that many

concepts of the classical Gromov-Witten theory such as the famous WDVV equations can be carried over to the tropical world. We use this to extend Mikhalkin's results to a certain class of gravitational descendants, i.e. we show that many of the classical gravitational descendants of $\mathbb{P}^2$ and $\mathbb{P}^1 \times \mathbb{P}^1$ can be computed by counting tropical curves satisfying certain incidence conditions and with prescribed valences of their vertices. Moreover, the presented theory is not restricted to plane curves and therefore provides an important tool to derive similar results in higher dimensions.

A more detailed chapter synopsis can be found at the beginning of each individual chapter.

This thesis contains material from my (partly published) articles [AR07], [MR08], [AR08] and [R08]. In particular, it contains joint work with Lars Allermann and Hannah Markwig. The contributions of each article are listed at the beginning of each chapter, as well.

Financial support

Financial support was provided by the Konrad-Adenauer-Stiftung via a graduate fellowship. I would like to thank the Konrad-Adenauer-Stiftung for its financial support (also for several stays abroad), the enjoyable seminars and for meeting so many funny and refreshing people.

Danksagung

Ich danke meinem Betreuer Andreas Gathmann — für die vielen Ideen und Anregungen, für die Hilfe bei Problemen, für die offene und entspannte Arbeitsatmosphäre.

Ich danke Hannah Markwig — für die Einladung nach Michigan und die gemütliche Unterbringung bei ihr zuhause.

Ich danke Henning Meyer — fürs Korrekturlesen.

Ich danke Jessica Rigden und Stefan Steidel — für Frühstück, Kaffee und Schwätzchen.

Ich danke Michael Kerber — für die gemeinsamen Gespräche über Gott und die Welt, über Politik und Frauen (also ich über Politik, er über Frauen), für den ersten Satz dieser Arbeit, in Erwartung einer weiteren Einladung zum Bier!

Ich danke meiner Familie, ganz besonders meinen Eltern — für die Unterstützung nach allen Kräften und mit viel Liebe.

Ich danke meinem Bruder Sebastian — fürs gemeinsame Musik machen und für die Einladung nach Norwegen.

Ich danke Andrea Wolf — für die geleisteten Überstunden, als die Zeit knapp wurde.

Ich danke meiner Tochter und "Herzkersch" Nora — für die vielen schönen und fröhlichen gemeinsamen Stunden, für ihre erfrischende Sicht der Dinge, fürs Flöte spielen.

Ich danke Christian Eder — einem guten Freund.

1 Tropical intersection theory

Introduction

Right from the beginning, in the early tropical works on enumerative questions, the need for a powerful tropical intersection theory was vivid. For example, in order to show that the number of rational plane tropical curves through generic points does not depend on the point configuration, a tedious and careful case by case study was necessary in [GM05]. In contrast to this, in the classical algebro-geometric setting such invariance statements usually follow automatically from the use of intersection theory and rational equivalence. More general, the algebro-geometric intersection theory is the indispensable basis of modern enumerative geometry — the main objects, Gromov-Witten invariants and gravitational descendants, cannot even be defined without intersection theory. For these reasons, it is an urgent task to make a tropical intersection theory available that makes tropical geometry more successful with regard to enumerative questions. Therefore the first chapter of this thesis rigorously establishes a useful and general tropical intersection theory.

The first step of this undertaking is to make clear what tropical varieties actually are! Note that the early works of tropical geometry did not even agree on what a tropical curve exactly is. So, first of all, we deal with polyhedral complexes, the balancing condition and the precise definition of a tropical variety, or tropical cycle, as we usually call it (cf. section 1.1).

Then we define rational functions and construct their "locus of zeros and poles". This basic construction leads to a general intersection product between rational functions (or Cartier divisors) and tropical cycles, which satisfies properties such as commutativity, locality, and so on (cf. section 1.2). We then add morphisms to our framework, study the intersection-theoretic functors they define and prove the projection formula which relates these functors to the intersection product (cf. section 1.3). In view of applications in enumerative geometry, we then introduce a concept of rational equivalence and compute the corresponding Chow groups of a vector space (cf. section 1.4). Finally, we also define an intersection product between cycles contained in the same vector space and prove the expected properties like commutativity, associativity and compatibility with rational equivalence (cf. section 1.5).

Throughout the text, we compare these constructions to the corresponding ones in toric geometry. For example, we show that our intersection product of cycles is equivalent to the fan displacement rule for cohomology classes described in [FS94]. Moreover, we prove that, under a certain genericity condition, taking complete intersections (classically) commutes with the process of tropicalization (cf. section 1.6). This can be regarded as a natural extension of the Bernshtein bound on the number of solutions of a system of polynomial equations.

One should note that, in spite of the relationship to toric geometry, our approach is self-contained inside tropical geometry. The proofs are of a combinatorial kind and do usually not depend on classical statements from algebraic geometry.

This first chapter mainly emerged from the material published in [AR07], [AR08] and the first section of [R08]. As far as the first two articles are concerned, this is joint work with Lars Allermann, and it is very hard to single out the contributions each of us made. As far as it can be told, main ideas of Lars Allermann are contained in sections 1.1 and 1.3, whereas main ideas of mine are contained in sections 1.2 and 1.4. Section

1.5 contains important contributions of both of us. Moreover, I omit those parts which are to a large extent the work of Lars Allermann. The presentation of the material as well as section 1.6, dealing with the process of tropicalization, are completely new.

1.1 Tropical cycles

In this first section, we will study in detail the basic geometric objects of tropical geometry, tropical cycles. They are given by polyhedral complexes in a vector space which satisfy the well-known balancing condition. However, as the tropicalization of a classical variety usually only comes as a polyhedral set, without an explicit polyhedral complex fixed, a suitable definition of tropical cycles is necessary. In particular, any refinement of a balanced polyhedral complex describes the same tropical cycle. This causes some technical issues, with which we also deal here.

1.1.1 Polyhedral complexes

In the following, let Λ be a free $\mathbb{Z}$-module of finite rank and let $V := \mathbb{R} \otimes \Lambda$ be the associated finite-dimensional vector space. A *(non-empty rational convex) polyhedron* σ is a subset of V whose elements satisfy a finite set of given inequalities of the form $\lambda_i(x) \geq a_i$, where $\lambda \in \Lambda^{\vee}$ is an integer linear form and $a \in \mathbb{R}$ is a real constant. A *face* τ of a polyhedron σ is a subpolyhedron obtained by transforming some of the inequalities into equalities. Equivalently, a face is the locus of minimality of a linear form on σ. The notation is $\tau < \sigma$. The *linear* subspace generated by a polyhedron σ is denoted by V_{σ}, the corresponding lattice is $\Lambda_{\sigma} := V_{\sigma} \cap \Lambda$. The *dimension of* σ is the dimension of V_{σ} (or the rank of Λ_{σ}). The *relative interior* $\mathrm{RelInt}(\sigma)$ *of* σ is the complement of all proper faces of σ. Equivalently, $\mathrm{RelInt}(\sigma)$ is the topological interior of σ considered as a subset in the affine space spanned by σ. In particular, $\mathrm{RelInt}(\sigma)$ is never empty (if σ is not empty).

A *polyhedral complex* $\mathcal{X}$ is a finite set of polyhedra (also called the *cells* of $\mathcal{X}$) such that the following two conditions hold:

(a) Any face τ of a cell $\sigma \in \mathcal{X}$ is again contained in $\mathcal{X}$.

(b) For any pair of cells $\sigma_1, \sigma_2 \in \mathcal{X}$ the intersection $\sigma_1 \cap \sigma_2$ is a common face.

The *support* $|\mathcal{X}|$ *of* $\mathcal{X}$ is the union of all polyhedra in $\mathcal{X}$. In the following, all occurring polyhedral complexes are *pure-dimensional*, which means that all maximal cells have the same dimension. These top-dimensional cells are called *facets*, the codimension one cells are called *ridges*, cells of dimension one are called *edges* or *rays* and cells of dimension zero are called *vertices*. The set of all cells of a given dimension d is denoted by $\mathcal{X}^{(d)}$. A *fan* $\mathcal{F}$ is a polyhedral complex that contains the cell $\{0\}$ and whose cells are all cones.

Example 1.1.1 (a) The easiest example of a polyhedral complex is given by a single polyhedron together with all its faces. Even more special, if $W \subseteq V$ is a subspace (with rational slope), then the one element set $\{W\}$ forms a polyhedral complex, which we will simply denote by W by abuse of notation.

(b) For every non-zero integer affine form $\lambda(x) + a, \lambda \in \Lambda^\vee, a \in \mathbb{R}$ the polyhedral complex

$$\mathcal{H}_{(\lambda,a)} := \Big\{ \{x|\lambda(x) + a \geq 0\}, \{x|\lambda(x) + a = 0\}, \{x|\lambda(x) + a \leq 0\} \Big\}$$

subdivides V into two half-spaces.

(c) Let $u_1, \ldots, u_r$ be a basis of Λ and $u_0 := -u_1 - \cdots - u_r$ (if $\Lambda = \mathbb{Z}^r$, we choose the *negatives* of the standard basis). For every proper subset $I \subsetneq \{0, \ldots, r\}$ we form the cone τ_I generated by the vectors $u_i, i \in I$.

Now, for every $d \in \{0, \ldots, r\}$ we collect all the cones τ_I of dimension lower or equal than d (i.e. with $|I| \leq d$) in the set $\mathcal{L}_d^r$. As the faces of a cone τ_I are precisely the cones $\tau_J, J \subseteq I$ and as $\tau_{I_1} \cap \tau_{I_2} = \tau_{I_1 \cap I_2}$, $\mathcal{L}_d^r$ is in fact a polyhedral complex of pure dimension d. The following picture shows the case $r = 2$.

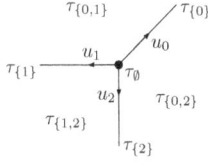

$\mathcal{L}_0^2$ is just a point, $\mathcal{L}_1^2$ will later describe a tropical line and $\mathcal{L}_2^2$ is just a subdivision of V.

(d) Let $\mathcal{X}$ and $\mathcal{Y}$ be two polyhedral complexes. Then we define the *intersection of $\mathcal{X}$ and $\mathcal{Y}$* to be

$$\mathcal{X} \cap \mathcal{Y} := \{\sigma \cap \sigma' | \sigma \in \mathcal{X}, \sigma' \in \mathcal{Y}\}.$$

It can easily be checked that $\mathcal{X} \cap \mathcal{Y}$ is also a polyhedral complex.

(e) Let $\mathcal{X}$ and $\mathcal{X}'$ be two polyhedral complexes in the vector spaces V resp. V'. Then it is obvious that the *cartesian product of $\mathcal{X}$ and $\mathcal{X}'$*

$$\mathcal{X} \times \mathcal{X}' := \{\tau \times \tau' | \tau \in \mathcal{X}, \tau' \in \mathcal{X}'\}$$

is also a polyhedral complex (in $V \times V'$) such that $|\mathcal{X} \times \mathcal{X}'| = |\mathcal{X}| \times |\mathcal{X}'|$. In the literature, this construction is often called "direct sum" (cf. [Zi94, definition 7.6]), which we avoid here since we are going to define another sum of polyhedral complexes later.

Let $\mathcal{X}$ and $\mathcal{X}'$ be two polyhedral complexes. We say $\mathcal{X}'$ is a *refinement of $\mathcal{X}$*, denoted by $\mathcal{X} \lhd \mathcal{X}'$, if $|\mathcal{X}| = |\mathcal{X}'|$ and if for all cells $\tau \in \mathcal{X}$ there exists a cell $\tau' \in \mathcal{X}'$ containing τ. Equivalently, every cell of $\mathcal{X}'$ is a union of cells from $\mathcal{X}$. For two polyhedral complexes $\mathcal{X}$ and $\mathcal{Y}$ with agreeing support $|\mathcal{X}| = |\mathcal{Y}|$, the intersection $\mathcal{X} \cap \mathcal{Y}$ is a common refinement.

Let us have a look at how nice refinements of an arbitrary fan can be. This will be quite useful in the following. A cone τ in $V = \Lambda \times \mathbb{R}$ is called *unimodular* if it can be generated by a part of a lattice basis. In other words, we find vectors $v_1, \ldots, v_d \in \Lambda$ such that $\dim(\tau) = d$, $\tau = \mathbb{R}_{\geq 0} v_1 + \ldots + \mathbb{R}_{\geq 0} v_d$ and $\Lambda_\tau = \mathbb{Z} v_1 + \ldots + \mathbb{Z} v_d$. In particular, any unimodular cone τ is *simplicial*, i.e. can be generated by $\dim(\tau)$ many vectors. A fan $\mathcal{F}$ is called *simplicial/unimodular* if all its cones are simplicial/unimodular.

Proposition 1.1.2 (Unimodular refinements)
Let $\mathcal{F}$ be a fan in $V = \Lambda \times \mathbb{R}$. Then there exists a unimodular fan $\mathcal{F}'$ which is a refinement of $\mathcal{F}$ and contains all unimodular cells τ of $\mathcal{F}$ (i.e. a cone in $\mathcal{X}$ which is already unimodular need not be refined).

Proof. This result is well-known in toric geometry where it guarantees the existence of a torus-equivariant resolution of singularities (cf. [Fu93, section 2.6]). As in [Fu93] the proof is only sketched, we give a complete argument here for the reader's convenience.

The basic construction is to refine $\mathcal{F}$ for a given vector $v \in |\mathcal{F}| \cap \Lambda$ such that the refinement contains the ray $\mathbb{R}_{\geq 0} v$. This works as follows: For any $\tau \in \mathcal{F}$ containing v, we remove τ from $\mathcal{F}$ and instead add the sums of $\mathbb{R}_{\geq 0} v$ with any face of τ not containing v.

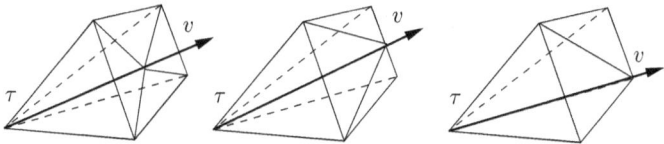

It is easy to check that this forms a polyhedral subdivision of τ and, as the construction agrees on faces of τ containing v, gives a refinement of

$\mathcal{F}$ in total.

First, we use this construction to make $\mathcal{F}$ simplicial. We choose a cone $\tau \in \mathcal{F}$ which is of minimal dimension while being *not* simplicial. This ensures that all faces of τ simplicial. Then we carry out the above refinement construction for a lattice vector $v \in \mathrm{RelInt}(\tau)$. This means the faces of τ are maintained while τ is replaced by cones of the form $\tau' + \mathbb{R}_{\geq 0}v, \tau' < \tau$, which are obviously simplicial cones. Inductively this procedure provides a simplicial refinement of $\mathcal{F}$.

Now we assume that $\mathcal{F}$ is simplicial. We can measure the non-unimodularity of a simplicial cone τ by the *index* $\mathrm{ind}(\tau) := [\Lambda_\tau : \mathbb{Z}v_1 + \ldots + \mathbb{Z}v_d]$, where the v_i are primitive lattice vectors generating τ. Take a $\tau \in \mathcal{F}$ with maximal index $\mathrm{ind}(\tau)$ (> 1 if $\mathcal{F}$ is not unimodular yet). It follows that there exists a non-zero primitive lattice vector $v \in \Lambda$ with

$$v = \frac{a_1}{b}v_1 + \ldots + \frac{a_d}{b}v_k \in \Lambda$$

for suitable integers $b \in \mathbb{Z} \setminus \{0\}, 0 \leq a_i < b$. We refine $\mathcal{F}$ along v and get new cones, for example $\tau_d = \mathbb{R}_{\geq 0}v_1 + \ldots + \mathbb{R}_{\geq 0}v_{d-1} + \mathbb{R}_{\geq 0}v$, if $a_d \neq 0$. Now it follows from

$$\mathbb{Z}v_1 + \ldots + \mathbb{Z}v_{d-1} + \mathbb{Z}a_d v_d = \mathbb{Z}v_1 + \ldots + \mathbb{Z}v_{d-1} + \mathbb{Z}bv$$

that $\mathrm{ind}(\tau_d) = (a_d/b)\,\mathrm{ind}(\tau) < \mathrm{ind}(\tau)$. As for any face τ' of τ_d we obviously have $\mathrm{ind}(\tau') \leq \mathrm{ind}(\tau_d)$, we can conclude that all "new" cones of our refinement have strictly lower index, and therefore by repeating the process we will eventually end up with a unimodular refinement. Note also that we have not refined any unimodular cone of $\mathcal{F}$ (as v can only lie in cones which are not unimodular). $\qquad\square$

1.1.2 Balancing condition

We now want to describe what makes a polyhedral complex "tropical". This is the so-called balancing condition, whose precise formulation needs the following preliminaries:

Whenever τ is a codimension one face of σ, we define the *primitive generator* $u_{\sigma/\tau}$ *of* σ *modulo* τ to be the element in Λ/Λ_τ with the following properties:

- $u_{\sigma/\tau}$ generates the ray $\bar{\sigma}$, i.e. $\bar{\sigma} = \mathbb{R}_{\geq 0} u_{\sigma/\tau} + \bar{\tau}$, where $\bar{\sigma}$ and $\bar{\tau}$ are the images of σ and τ under the quotient map $V \to V/V_\tau$.

- $u_{\sigma/\tau}$ is primitive, i.e. $\mathbb{Z} u_{\sigma/\tau} = \Lambda_\sigma/\Lambda_\tau$.

Representatives of $u_{\sigma/\tau}$ in V are usually called *primitive representatives of* σ *modulo* τ and denoted by $v_{\sigma/\tau}$. For a lattice vector $v_{\sigma/\tau} \in \Lambda$, this is equivalent to

- $v_{\sigma/\tau}$ points from τ towards σ, i.e. if λ is a linear form whose minimal locus on σ is τ, then $\lambda(v_{\sigma/\tau}) > 0$,

- $v_{\sigma/\tau}$ generates σ modulo τ primitively, i.e. $\mathbb{Z} v_{\sigma/\tau} + \Lambda_\tau = \Lambda_\sigma$.

A polyhedral complex is called *weighted* if it is equipped with a *weight function* $\omega_\mathcal{X} : \mathcal{X}^{(\dim(\mathcal{X}))} \to \mathbb{Z}$ that assigns an integer weight to every facet. Note that we allow weight 0 for notational reasons. In practice, only the *non-zero part* $\mathrm{NZ}(\mathcal{X}) := \{\tau | \tau < \sigma \text{ for some } \sigma \text{ with } \omega_\mathcal{X}(\sigma) \neq 0\}$ is the important part of $\mathcal{X}$. In this sense, we define the *support* $|\mathcal{X}|$ of a weighted polyhedral complex to be $|\mathcal{X}| := |\mathrm{NZ}(\mathcal{X})|$. Moreover, we extend the definition of refinements $\mathcal{X} \lhd \mathcal{X}'$ to possibly weighted complexes by replacing a weighted one by its non-zero part. In this situation, if $\mathcal{X}$ is weighted but $\mathcal{X}'$ is not, we get an induced weight function $\omega_{\mathcal{X}'}$ assigning to every facet σ of $\mathcal{X}'$ the weight of the unique facet of $\mathcal{X}$ containing σ.

Now we are ready to formulate the balancing condition:

Definition 1.1.3 (Balancing condition)

Let $\mathcal{X}$ be a weighted polyhedral complex of dimension d. We call $\mathcal{X}$ *balanced* if for every ridge $\tau \in \mathcal{X}^{(d-1)}$ the following *balancing condition* holds: The weighted sum of the primitive generators of all facets σ around τ vanishes (modulo V_τ), i.e.

$$\sum_{\substack{\sigma \in \mathcal{X}^{(d)} \\ \tau < \sigma}} \omega(\sigma) u_{\sigma/\tau} = 0 \in V/V_\tau,$$

or, in terms of primitive representatives:

$$\sum_{\substack{\sigma \in \mathcal{X}^{(d)} \\ \tau < \sigma}} \omega(\sigma) v_{\sigma/\tau} \in V_\tau$$

Example 1.1.4 (a) In the following, a subspace $W \subseteq V$ is always understood to be weighted with global weight $\omega(W) = 1$. As no ridge exists in this case, the balancing condition is trivially fulfilled and W is a balanced polyhedral complex.

(b) Let $\lambda(x) + a$ be an integer affine form and let $\mathcal{H}_{(\lambda,a)} = \{\sigma_\geq, \tau_=, \sigma_\leq\}$ be the associated half-space subdivision of V. In this case, a primitive representative $v_{\sigma_\geq/\tau_=}$ is a vector satisfying $\lambda(v_{\sigma_\geq/\tau_=}) = 1$ (resp. $\lambda(v_{\sigma_\leq/\tau_=}) = -1$). Hence the two primitive generators are opposite to each other and if we use the weights $\omega(\sigma_\geq) = \omega(\sigma_\leq) = 1$, $\mathcal{H}_{(\lambda,a)}$ is balanced. However, this balanced complex is going to be identified with V in the next subsection.

(c) Recall our polyhedral complexes $\mathcal{L}_d^r$ from example 1.1.1 (c). Again we weight every facet $\tau_I, |I| = d$ with $\omega(\sigma_I) = 1$. Note that if τ_J is a ridge of τ_I (i.e. $I \setminus J = \{k\}$), then a primitive representative of τ_I modulo τ_J is given by u_k. So the weighted sum around τ_J equals $\sum_{k \notin J} u_k = -\sum_{k \in J} u_k \in V_{\tau_J}$ and therefore $\mathcal{L}_d^r$ is also balanced. The following picture illustrates the one-dimensional examples $\mathcal{L}_1^1$, $\mathcal{L}_1^2$ and $\mathcal{L}_1^3$, where the balancing condition is satisfied around $\{0\}$.

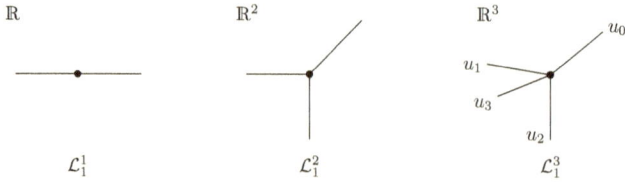

(d) Let $\mathcal{X}$ and $\mathcal{X}'$ be two balanced polyhedral complexes in the vector spaces V resp. V'. A facet of $\mathcal{X} \times \mathcal{X}'$ is given as the cartesian product $\sigma \times \sigma'$ of two facets of $\mathcal{X}$ and $\mathcal{X}'$ respectively and we can define the weight function

$$\omega_{\mathcal{X} \times \mathcal{X}'}(\sigma \times \sigma') = \omega_{\mathcal{X}}(\sigma) \cdot \omega_{\mathcal{X}'}(\sigma').$$

Then $\mathcal{X} \times \mathcal{X}'$ is also balanced, as a ridge of $\mathcal{X} \times \mathcal{X}'$ is given by $\tau \times \sigma'$ or $\sigma \times \tau'$ (dimensions are understood) and the sum of weighted primitive generators equals $\omega_{\mathcal{X}'}(\sigma')$ (resp. $\omega_{\mathcal{X}}(\sigma)$) times the sum around τ (resp. τ').

1.1.3 Tropical cycles

As we will see in section 1.6, the tropicalization of an algebraic variety equals the support of balanced polyhedral complexes. However, in general the tropicalized variety does not fix one specific polyhedral complex, but only the underlying set. Let us make the appropriate definitions.

A *(pure-dimensional) polyhedral set* X *in* V is the union of finitely many polyhedra $X = \sigma_1 \cup \ldots \cup \sigma_n \subseteq V$ of equal dimension. Obviously, the support of a polyhedral complex is a polyhedral set. The other way around, a (weighted) polyhedral complex $\mathcal{X}$ whose (weighted) support equals X is called a *(weighted) polyhedral structure of X*.

Lemma 1.1.5

Every polyhedral set X admits a polyhedral structure $\mathcal{X}$.

Proof. Assume $X = \sigma_1 \cup \ldots \cup \sigma_n$, where each polyhedron σ_i is given by a finite set E_i of inequalities , and let $E := \bigcup_{i=1}^{n} E_i$ be the union of all these inequalities. For each inequality $\lambda(x) + a \geq 0$ in E, we consider the polyhedral complex $\mathcal{H}_{(\lambda,a)}$ which subdivides V into two affine half-spaces (cf. example 1.1.1 (b)). Now we form the intersection of all these complexes $\mathcal{H}_{(\lambda,a)}$, where (λ, a) runs through E, obtaining a new polyhedral complex $\mathcal{H}$. Obviously, every σ_i is a union of cells in $\mathcal{H}$ and therefore

$$\mathcal{X} := \{\tau \in \mathcal{H} | \tau \subseteq X\}$$

is a polyhedral structure of X. $\qquad \square$

Remark 1.1.6

In fact, this lemma is not very important as in the following most polyhedral sets will come as the support of a polyhedral complex. However, the construction used in the proof will be useful several times (e.g. in the next remark). One should also note that in general a polyhedral set does not have a well-defined minimal or otherwise canonical polyhedral structure. The following picture illustrates the problem, even though the displayed polyhedral set is not pure-dimensional.

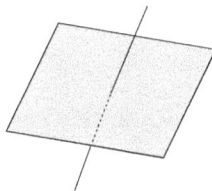

A pure-dimensional example is given by two 2-planes in $\mathbb{R}^4$ which intersect in a point.

Remark 1.1.7

Let Y and X be two polyhedral sets such that $Y \subseteq X$. Then the construction of the previous lemma applied to the polyhedral set "$X \cup Y$" shows that there exist polyhedral structures $\mathcal{Y}$ of Y and $\mathcal{X}$ of X such that $\mathcal{Y} \subseteq \mathcal{X}$.

Now we are ready to define the basic geometric objects, namely tropical cycles.

Definition 1.1.8 (Tropical cycles)

A *tropical cycle X in V* is a polyhedral set together with a balanced weighted polyhedral structure $\mathcal{X}$. We identify two such structures $\mathcal{X}$ and $\mathcal{Y}$ if the induced weight functions on the common refinement $\mathcal{X} \cap \mathcal{Y}$ agree. To avoid conflicts, we use the notation $|X|$ when we want to consider the polyhedral set *without* the additional structure (and call it the *support of X*, as well).

A *tropical fan F* is a tropical cycle whose support $|F|$ is a cone in the general meaning that for every vector $v \in |F|$ the whole ray $\mathbb{R}_{\geq} v$ is contained in $|F|$.

Before we make some remarks concerning this definition, we prove a technical lemma which enlarges our available choice of polyhedral structures.

Lemma 1.1.9

Let X be a tropical cycle of dimension d. Then for every unweighted polyhedral complex $\mathcal{Y}$ of dimension d with $|X| \subseteq |\mathcal{Y}|$ there exists a canonical weight function on $\mathcal{Y}$ (possibly with zeros) such that $\mathcal{Y}$ becomes a balanced weighted polyhedral structure of X.

Proof. Let $\mathcal{X}$ be a balanced weighted polyhedral structure for X. Applying the construction of lemma 1.1.5 to the polyhedral set

$$|\mathcal{Y}| = \bigcup_{\tau \in \mathcal{X} \cup \mathcal{Y}} \tau,$$

we obtain a refinement $\mathcal{Y}'$ of $\mathcal{Y}$ with induced weight function

$$\omega_{\mathcal{Y}'}(\sigma) = \begin{cases} 0 & \text{if } \sigma \not\subseteq |X|, \\ \omega_{\mathcal{X}}(\sigma') & \text{if } \sigma \subseteq |X|, \end{cases}$$

where in the second case σ' is the unique facet of $\mathcal{X}$ containing σ. Obviously this definition makes $\mathcal{Y}'$ into a balanced weighted polyhedral structure of X. We carry this over to $\mathcal{Y}$ by defining the weight of a facet σ of $\mathcal{Y}$ to be $\omega_{\mathcal{Y}}(\sigma) := \omega_{\mathcal{Y}'}(\sigma')$, where σ' is a facet of $\mathcal{Y}'$ contained in σ. It remains to check that, if we choose another facet $\sigma'' \in \mathcal{Y}'$ contained in σ, the weights agree. This follows from the fact that two such facets are connected along other facets in σ via ridges intersecting the interior of σ (as RelInt(σ) cannot become disconnected by removing a polyhedral set of codimension 2). Thus we can assume that σ' and σ'' intersect in a ridge $\tau \in \mathcal{C}'$ which intersects the interior of σ. But this implies that σ' and σ'' are the only facets in $\mathcal{Y}'$ containing τ and that their primitive generators have opposite directions. Then the balancing condition ensures that the weights of σ and σ' agree. Therefore our definition of the weight function $\omega_{\mathcal{Y}}$ is well-defined and features the desired properties. □

Remark 1.1.10 (a) The previous lemma justifies that in the following we call any such $\mathcal{Y}$ with induced weight function a polyhedral structure of X (omitting "balanced weighted").

(b) For most of the following constructions with tropical cycles, the choice of a polyhedral structure $\mathcal{X}$ of X is necessary. Note that, in order to show that the construction does not depend on this choice, it suffices to show this for refinements of $\mathcal{X}$ (with induced weight function). This is because any two polyhedral structures of X admit a common refinement.

(c) Let X be a tropical cycle. X is a tropical fan as defined above if and only if the polyhedral structure of X can be chosen to be a fan (which is then called a *fan structure of X*). To see this, let X be a tropical fan and therefore $|X|$ a (general) cone given as the union of polyhedra $|X| = \sigma_1 \cup \ldots \cup \sigma_n$. Then every polyhedron σ_i can be replaced by the cone spanned by σ_i over 0. Thus, we can assume that all σ_i are cones. In this case, the construction of lemma 1.1.5 provides a polyhedral structure $\mathcal{X}$ of X only containing cones. We can also assume that $\{0\}$ is contained in $\mathcal{X}$ (if not, we intersect with an arbitrary complete fan). Thus $\mathcal{X}$ is a fan as claimed.

Example 1.1.11

All weighted complexes occurring in example 1.1.4 are balanced and therefore define tropical cycles. However, we can use different polyhedral structures in all cases as well.

- In (a) and (b), two different polyhedral structures of the polyhedral set V are given, where the latter one is in general *not* a fan structure. In fact, a third polyhedral structure is given by $\mathcal{L}_r^r$ from (c). The associated tropical fan is denoted by V (resp. W in (a)) as well.

- The tropical fans defined by the balanced polyhedral complexes $\mathcal{L}_d^r$ in (c) are denoted by L_d^r and called *(degenerated) tropically linear spaces*.

- Let X and X' be tropical cycles in V and V' respectively and choose arbitrary polyhedral structures $\mathcal{X}$ and $\mathcal{X}'$. In (d) we defined the balanced polyhedral complex $\mathcal{X} \times \mathcal{X}'$. The associated tropical cycle is denoted by $X \times Y$ and is called the *cartesian product of X and*

Y. If we choose refinements of $\mathcal{X}$ and $\mathcal{X}'$ instead, the construction provides an refinement of $\mathcal{X} \times \mathcal{X}'$ with compatible weights, hence the previous remark shows that the definition of $X \times Y$ does not depend on the choice of polyhedral structures.

There are some reasons why tropical cycles are called cycles and not varieties.

Contra variety: Not every tropical cycle can be obtained as the tropicalization of an algebraic variety. In particular, via valuation no negative weights can occur. We will deal with this more precisely in section 1.6. Therefore one might like to give the term "variety" a more restrictive meaning, e.g. requiring non-negative weights.

Pro cycle: As we will see in the following, tropical cycles are not only geometric objects but also the basic elements for intersection theory, as the classical cycles. A first example might be the following construction of the sum of two tropical cycles.

Definition 1.1.12 (Sums of cycles, subcycles)
Let X and Y be two tropical cycles in V. By lemma 1.1.9, a polyhedral structure $\mathcal{Z}$ of the union $|X| \cup |Y|$ carries the two weight functions ω_X induced by X and ω_Y induced by Y. Obviously, the sum $\omega_X + \omega_Y$ still satisfies the balancing condition. We define the *sum* $X + Y$ to the tropical cycle determined by $\mathcal{Z}$ with weight function $\omega_X + \omega_Y$. Note that the support of $X + Y$ is in general only a subset of $|X| \cup |Y|$, as some weights might add up to zero.

We call a tropical cycle Y a *subcycle of* X if $|Y| \subseteq |X|$. Note that by remark 1.1.7 there exist polyhedral structures $\mathcal{Y}$ of Y and $\mathcal{X}$ of X such that $\mathcal{Y} \subseteq \mathcal{X}$. The set of d-dimensional tropical cycles contained in a given cycle X is denoted by $Z_d(X)$ and forms a group with respect to the sum constructed above (we always include $\emptyset$ as neutral element). In particular, $Z_d(V)$ is the *group of all d-dimensional cycles in* V.

Example 1.1.13

For all $d < e \leq r$, the tropical fan L_d^r is a subcycle of L_e^r.

1.1.4 Minkowski weights

Let us now describe in which sense tropical cycles appear in *toric* geometry. The main reference for this is [FS94].

Definition 1.1.14 (Ω-directional cycles)

Let Ω be a complete fan in V (i.e. $|\Omega| = V$). A tropical fan F is called Ω-*directional* if $|F| \subseteq |\Omega^{(\dim F)}|$. In this case, by lemma 1.1.9 we get an induced weight function on $\Omega^{(\dim F)}$ such that $\Omega^{(\leq \dim F)}$ is a polyhedral structure of F. We denote by $Z_d(\Omega) := Z_d(|\Omega^{(d)}|$ the group of all d-dimensional Ω-directional tropical fans.

Now let us fix a complete fan Ω in V and let $\mathbf{X} := \mathbf{X}(\Omega)$ be the associated compact toric variety. In [FS94, section 2], the authors introduce d-dimensional *Minkowski weights*, which are weight functions ω on $\Omega^{(d)}$ satisfying for any ridge $\tau \in \Omega^{(d-1)}$ and any linear form $\lambda \in \Lambda_\tau^\perp$ the equation

$$\sum_{\substack{\sigma \in \Omega^{(d)} \\ \tau < \sigma}} \omega(\sigma)\lambda(u_{\sigma/\tau}) = 0. \tag{1.1}$$

Of course, this equation is satisfied for all $\lambda \in \Lambda_\tau^\perp$ if and only if

$$\sum_{\substack{\sigma \in \Omega^{(d)} \\ \tau < \sigma}} \omega(\sigma)u_{\sigma/\tau} = 0 \in V/V_\tau,$$

which is precisely our well-known balancing condition. Thus the Minkowski weight ω defines an Ω-directional tropical cycle $X_\omega \in Z_k(\Omega)$. The other way around, every Ω-directional cycle defines a Minkowski weight. Therefore we can reformulate [FS94, theorem 2.1] as follows:

Theorem 1.1.15 (Tropical cycles and toric cohomology classes)
Let Ω be a complete fan in V of dimension r. The Chow cohomology group $A^{r-d}(\mathbf{X})$ of the toric variety $\mathbf{X} = \mathbf{X}(\Omega)$ is canonically isomorphic to the group of d-dimensional tropical cycles contained in $|\Omega^{(d)}|$ (for all d), i.e.

$$A^{r-d}(\mathbf{X}) \cong Z_d(\Omega).$$

We will often take the opposite point of view and fix a tropical fan F instead of Ω. Then the following reformulation is more appropriate:

Corollary 1.1.16
Let X be a tropical fan of dimension d. Then for any complete fan Ω with $|X| \subseteq |\Omega^{(d)}|$, X induces a cohomology class of $\mathbf{X} := \mathbf{X}(\Omega)$ of dimension d, denoted by $\gamma_X \in A^{r-d}(\mathbf{X})$.

Let γ be a cohomology class of $\mathbf{X}$ of codimension $r-d$. The isomorphism of theorem 1.1.15 is given by the weight function

$$\omega_\gamma : \Omega^{(d)} \to \mathbb{Z},$$
$$\sigma \mapsto \deg(\gamma \cap [V(\sigma)]),$$

where $V(\sigma)$ denotes the closure of the orbit associated to σ in $\mathbf{X}$. In other words, the weight of the facet σ in the tropical fan X_γ corresponding to γ is given by the degree of the intersection of γ with the orbit closure $V(\sigma)$. Theorem 1.1.15 then follows from the fact that $A_*(\mathbf{X})$ is generated by the orbit closures $V(\sigma)$ and that the Kronecker duality homomorphism

$$\mathcal{D}_{\mathbf{X}} : A^k(\mathbf{X}) \to \operatorname{Hom}(A_k(\mathbf{X}), \mathbb{Z})$$
$$\gamma \mapsto (\alpha \mapsto \deg(\gamma \cap \alpha))$$

is an isomorphism in the case of complete toric varieties. The balancing condition of X_γ can be checked as follows. Let τ be a cell of dimension

$d - 1$ and $\lambda \in \Lambda_\tau^\perp$. Then λ defines a rational function x^λ on the toric variety $V(\tau)$ and its associated Weil divisor (cf. [FS94, section 3.3]) is

$$\operatorname{div}(x^\lambda) = \sum_{\substack{\sigma \in \Omega^{(d)} \\ \tau < \sigma}} \lambda(u_{\sigma/\tau}) \cdot [V(\sigma)].$$

This expression gives a relation on $A_{r-d}(\mathbf{X})$ and therefore must be zero when capped with γ. This precisely gives the balancing condition 1.1 mentioned at the beginning of this subsection.

1.2 Cartier and Weil divisors

The goal of this section is to define rational functions and compute their Weil divisors of zeros and poles, i.e. a tropical cycle of codimension 1 that reflects the behaviour of the function. This leads to a general intersection product of rational functions/Cartier divisors and tropical cycles. We also relate this intersection product to the cup-product of a cohomology class and a Cartier divisor on a toric variety. Finally, we have a look at convex functions, the tropical version of regular functions, and irreducible cycles.

1.2.1 Rational functions

In the previous section we defined the basic geometric objects, namely tropical cycles. We now move forward by equipping these geometric objects with an appropriate class of functions. This class is basically given by "piecewise (affine) linear" functions.

First, to make things simpler in the following, let us specify a polyhedral kind of open sets. Let X be a tropical cycle in V. An open set $U \subseteq |X|$ is called *polyhedral open* if, for a suitable polyhedral structure $\mathcal{X}$, U is the union of the relative interiors of a choice of cells $S \subseteq \mathcal{X}$, i.e.

$$U = \bigcup_{\tau \in S} \operatorname{RelInt}(\tau).$$

For any $\tau \in \mathcal{X}$ we define the *(polyhedral open) neighbourhood of τ* by

$$U(\tau) := \bigcup_{\substack{\sigma \in \mathcal{X} \\ \tau < \sigma}} \operatorname{RelInt}(\sigma).$$

Note that $U(\tau)$ does not contain τ but its relative interior. A set U is polyhedral open if and only if for any $\tau \in \mathcal{X}$ with $\operatorname{RelInt}(\tau) \subseteq U$, the neighbourhood $U(\tau)$ is contained in U (for a fine enough polyhedral structure $\mathcal{X}$). Finite intersections and unions of polyhedral open sets are again polyhedral open, as the involved polyhedral structures have a common refinement. Of course, the polyhedral open sets (still) generate the usual euclidean topology of $|X|$.

Definition 1.2.1 (Rational functions)
Let X be a tropical cycle in V and let $U \subseteq |X|$ be a polyhedral open set. A function $\varphi : U \to \mathbb{R}$ is called a *rational function on U* if there exists a polyhedral structure of X providing U as union of relative interiors and such that φ is integer affine on each cell τ intersecting U. Here, *integer affine* means that $\varphi|_\tau$ is the sum of an integer linear form $\lambda \in \Lambda_\tau^\vee$ and a real constant. This λ is uniquely fixed by φ and is called the *linear part of φ on τ*, also denoted by $\varphi_\tau := \lambda$. The polyhedral set of points in U where φ is *not* locally affine is called the *support of φ* and denoted by $|\varphi|$. The group (with respect to addition) of all rational functions on U is denoted by $\operatorname{Rat}(U)$ (where we omit the absolute value bars if $U = |X|$).

A rational function φ on a tropical fan F is called a *fan function* if it is linear on any ray $\mathbb{R}_{\geq 0} v, v \in |X|$. Equivalently, $\varphi(0) = 0$ and the polyhedral structure on whose cells φ is affine (resp. linear) can be chosen to be a fan.

Example 1.2.2 (a) Let X be a subcycle of Y. Then the restriction of any rational function on Y to X is again a rational function (cf. next remark).

(b) Every affine form $\lambda(v) + a$ is a rational function on V (and every subcycle). If $a = 0$, it is a fan function.

(c) Let φ, ψ be two rational functions on X. Then $\max\{\varphi, \psi\}$ and $\min\{\varphi, \psi\}$ are rational functions again. Indeed, if $\mathcal{X}$ is a polyhedral structure of X on whose cells φ and ψ are affine, then for every cell $\tau \in \mathcal{X}$ the equation $\varphi = \psi$ can be described by an affine form α_τ. Take the intersection of $\mathcal{X}$ with all the half-space subdivisions $\mathcal{H}_{\alpha_\tau}$, then $\max\{\varphi, \psi\}$ and $\min\{\varphi, \psi\}$ are affine on the cells of this refinement of $\mathcal{X}$. This can be repeated to show that $\max\{\varphi_1, \ldots, \varphi_l\}$ and $\min\{\varphi_1, \ldots, \varphi_l\}$ are rational functions as well.

(d) Let $S \subseteq \Lambda^\vee$ be a finite choice of linear forms and let $a_\lambda, \lambda \in S$ be some real constants. This defines a *tropical Laurent polynomial* $f : V \to \mathbb{R}$ by

$$f(v) = -\min_{\lambda \in S}\{\lambda(v) + a_\lambda\}.$$

The choice of $-\min$ (instead of max or min) is due to the compatibility with the valuation approach (cf. section 1.6).

(e) Let Ω be a unimodular fan. Then the assignment of a value v_ϱ to the primitive generator of each ray $\varrho \in \Omega^{(1)}$ induces a rational function on $|\Omega|$ by linear extension on all cells of Ω. In particular, we define φ_ϱ to be the unique rational function which satisfies $\varphi_\varrho(u_{\varrho/\{0\}}) = 1$ and is identically zero on all other rays. Obviously, these functions form a basis for all fan functions which are linear on the cones of Ω.

Remark 1.2.3

Let us note an important difference to the case of classical rational functions on algebraic varieties. Our functions are by definition tropically
non-zero as the "zero" element of tropical arithmetics $-\infty$ is not even
contained in the value set of our functions. Even if we changed our definition such that the value $-\infty$ is allowed, this would not really make
sense. At best, we would add the constant $-\infty$-function (on connected
components). This has a strange consequence: A tropical rational function *cannot get "zero"* when restricted to a subcycle. Instead, for any
subcycle the restriction is again a "non-zero" rational function according
to the definition. As we will see later, this makes sure that we can define
intersection products *without* passing to classes of rational equivalence,
unlike in the classical case. But first, in order to construct intersection
products one needs to know what the locus of zeros and poles of a rational
function is, and this seems to be difficult if "zero" is not attained at all.
We will see in the following how this can be managed.

Again, let us mention the connections to toric geometry (see for example
[Fu93, section 3.4]):

Theorem 1.2.4 (Rational functions and toric Cartier divisors)
Let $\mathbf{X}$ *be the toric variety induced by a fan* Ω. *Then there is a canonical
isomorphism between the group of torus-equivariant Cartier divisors of* $\mathbf{X}$
and the group of fan functions φ *on* Ω *(i.e.* φ *is linear on every cone*
$\sigma \in \Omega$, *denoted by* $\varphi \in \mathrm{Rat}(\Omega)$*),*

$$\mathrm{Div}_T(\mathbf{X}) \cong \mathrm{Rat}(\Omega).$$

This isomorphism works as follows. Let φ be a fan function linear on
the cones of Ω. For every $\tau \in \Omega$ we choose an extension of $\lambda_\tau \in \Lambda^\vee$ of
φ_τ. Each such λ_τ defines a rational function x^{λ_τ} on $\mathbf{X}$ and therefore also
on the affine open subset $U_\tau = \mathrm{Spec}(K[\tau^\vee])$ of $\mathbf{X}$. We form the Cartier

divisor

$$[\{(U_\tau, x^{\lambda_\tau}), \tau \in \Omega\}],$$

and this is well-defined and does not depend on the choice of the extension λ_τ. This follows from the fact that the differences of different choices as well as the differences on the overlaps $U_\tau = U_{\sigma_1} \cap U_{\sigma_2}$ with $\tau = \sigma_1 \cap \sigma_2$ are given by rational functions $x^\lambda, \lambda \in \tau^\perp$, which are invertible in $K[\tau^\vee]$ and thus regular invertible on U_τ. Note that in comparison to [Fu93, section 3.4], our mapping between tropical rational functions and classical Cartier divisors differs by a minus sign, i.e. the rational function ψ_D associated to a torus equivariant Cartier divisor in [Fu93, section 3.4] is the negative of ours (cf. the minus sign in our definition of a tropical polynomial).

1.2.2 The zeros and poles of a rational function

Now we turn our attention towards the locus of zeros and poles of a given rational function. As mentioned above, this is not straightforward because the respective values $-\infty$ and ∞ are not attained by our functions. A first idea how to deal with this is given by Kapranov's theorem [EKL04, theorem 2.1.1]. It considers the case where the rational function is a tropical polynomial f and proves that the zero locus of this function is its locus of non-linearity $|f|$, together with weights that measure the change of slope of f at this locus. This can be visualized by considering the graph of the function f in $V \times \mathbb{R}$, which gives us another idea how to define the zeros (or poles) of a rational function: The graph of a rational function is usually not a tropical cycle itself as at the locus of non-linearity the balancing condition might fail. But this deficiency can be removed by adding some cells to the graph in a canonical way, and this procedure reveals the zeros and poles of the rational function. Let us make this precise.

Construction 1.2.5 (Zeros and poles of a rational function)

Let X be a tropical cycle of dimension d and let φ be a rational function on X. We fix a polyhedral structure $\mathcal{X}$ on whose cells φ is affine and denote by $\tilde{\tau}$ the polyhedron in $V \times \mathbb{R}$ obtained as the graph of $\varphi|_\tau$, i.e. $\tilde{\tau} = \{(v, \varphi(v)) | v \in \tau\}$. Then a polyhedral structure of the graph of φ in $|X| \times \mathbb{R}$ is given by $\tilde{\mathcal{X}} = \{\tilde{\tau} | \tau \in \mathcal{X}\}$, and this polyhedral complex can be canonically weighted via $\tilde{\omega}(\tilde{\sigma}) := \omega(\sigma)$.

Now let us check the balancing condition around a ridge $\tilde{\tau}$ of $\tilde{\mathcal{X}}$. Let σ be a facet containing τ and let $v_{\sigma/\tau} \in \Lambda$ denote a primitive representative. Then, by definition, $(v_{\sigma/\tau}, \varphi_\sigma(v_{\sigma/\tau})) \in \Lambda \times \mathbb{Z}$ is a primitive representative of $\tilde{\sigma}$ modulo $\tilde{\tau}$. Therefore the weighted sum of these primitive representatives around $\tilde{\tau}$ gives

$$\sum_{\substack{\tilde{\sigma} \in \tilde{\mathcal{X}} \\ \tilde{\tau} < \tilde{\sigma}}} \tilde{\omega}(\tilde{\sigma})(v_{\sigma/\tau}, \varphi_\sigma(v_{\sigma/\tau})) = \left(\sum_{\substack{\sigma \in \mathcal{X} \\ \tau < \sigma}} \omega(\sigma) v_{\sigma/\tau}, \sum_{\substack{\sigma \in \mathcal{X} \\ \tau < \sigma}} \omega(\sigma) \varphi_\sigma(v_{\sigma/\tau}) \right). \quad (1.2)$$

From the balancing condition of $\mathcal{X}$ it follows that the vector

$$v := \sum_{\substack{\sigma \in \mathcal{X} \\ \tau < \sigma}} \omega(\sigma) v_{\sigma/\tau} \in V_\tau$$

lies in V_τ and therefore $(v, \varphi_\tau(v))$ lies in $V_{\tilde{\tau}}$. We conclude that modulo $V_{\tilde{\tau}}$ our sum 1.2 equals

$$\left(0, \sum_{\substack{\sigma \in \mathcal{X} \\ \tau < \sigma}} \omega(\sigma) \varphi_\sigma(v_{\sigma/\tau}) - \varphi_\tau(v) \right) \in V \times \mathbb{R}.$$

So our first observation is that, due to the fact that in general φ is not affine (locally around $\tilde{\tau}$), the balancing condition fails at $\tilde{\tau}$. On the other hand, it is easy to see how we can make $\tilde{\mathcal{X}}$ balanced. We add the cone

$\eta(\tau) := \tilde{\tau} + (\{0\} \times \mathbb{R}_{\leq 0})$ with weight

$$\omega(\eta(\tau)) = \sum_{\substack{\sigma \in \mathcal{X} \\ \tau < \sigma}} \omega(\sigma) \varphi_\sigma(v_{\sigma/\tau}) - \varphi_\tau \Big(\sum_{\substack{\sigma \in \mathcal{X} \\ \tau < \sigma}} \omega(\sigma) v_{\sigma/\tau} \Big). \qquad (1.3)$$

As obviously $(0, -1) \in V \times \mathbb{R}$ is a primitive representative of $\eta(\tau)$ modulo $\tilde{\tau}$, the above calculation shows that now the balancing around $\tilde{\tau}$ holds.

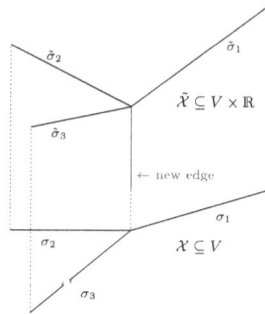

We can do this for all ridges, i.e. we can form the polyhedral complex

$$\bar{\mathcal{X}} := \tilde{\mathcal{X}} \cup \big\{ \eta(\tau) | \tau \in \tilde{\mathcal{X}} \setminus \tilde{\mathcal{X}}^{(d)} \big\}$$

and define the weights of the new facets by formula 1.3. $\bar{\mathcal{X}}$ can be regarded as the tropical closure of $\tilde{\mathcal{X}}$, even though such a notion does not exist in general. Now "intersection with $V \times \{-\infty\}$" (to be rigorized) provides us with a weighted polyhedral subcomplex of $\mathcal{X}$ of codimension 1 which is a good candidate for the locus of zeros and poles of φ. Here "intersection with $V \times \{-\infty\}$" means that for each $\eta(\tau)$ we get back τ, but with weight $\omega(\eta(\tau))$ if $\tau \in \mathcal{X}^{(d-1)}$.

Note that it is neither necessary nor helpful to intersect with $V \times \{+\infty\}$ as well. Instead, our above construction also measures the poles of φ by assigning negative weights to the respective cells. However, our construction stills lacks something: Note that $\bar{\mathcal{X}}$ also contains *new ridges* $\eta(\tau), \tau \in \mathcal{X}^{(d-2)}$ and it is not obvious that $\bar{\mathcal{X}}$ is also balanced around these new ones. This is equivalent to the question if the "intersection with $V \times \{-\infty\}$" provides a balanced complex. We will deal with this problem in the following. But let us first transform this discussion into an excact definition.

Definition 1.2.6 (Weil divisors of rational functions)
Let X be a tropical d-cycle and let φ be a rational function on X. We fix a polyhedral structure $\mathcal{X}$ on whose cells φ is affine. Then we define the weighted polyhedral complex $\varphi \cdot \mathcal{X}$ to be the $(d-1)$-skeleton of $\mathcal{X}$

$$\mathcal{X} \setminus \mathcal{X}^{(d)}$$

with weight function

$$\omega_\varphi : \mathcal{X}^{(d-1)} \to \mathbb{Z},$$
$$\tau \mapsto \sum_{\substack{\sigma \in \mathcal{X} \\ \tau < \sigma}} \omega(\sigma)\varphi_\sigma(v_{\sigma/\tau}) - \varphi_\tau\Big(\sum_{\substack{\sigma \in X \\ \tau < \sigma}} \omega(\sigma)v_{\sigma/\tau}\Big), \qquad (1.4)$$

where the $v_{\sigma/\tau}$ are arbitrary primitive representatives. The associated tropical cycle is called the *Weil divisor of* φ and is denoted by $\mathrm{div}(\varphi) = \varphi \cdot X$.

Remark 1.2.7
Let us make some remarks here.

- The weight function ω_φ is independent of the choice of primitive representatives, as a different choice only differs by elements in V_τ whose contributions cancel out.

- Due to the fact that the weight formula might take the value 0, we have $|\mathrm{div}(\varphi)| \subset |\mathcal{X} \setminus \mathcal{X}^{(d)}|$ but not $|\mathrm{div}(\varphi)| = |\mathcal{X} \setminus \mathcal{X}^{(d)}|$, in general.

- If φ is globally affine, its divisor will be the zero-cycle $\emptyset$, as in this case we can permute the sum and φ_σ in the first part of the weight formula 1.4.

- Our construction is local: The weight $w_\varphi(\tau)$ only depends on the behaviour of φ in the neighbourhood $U(\tau)$ of τ. This will be made more precise in the next subsection.

- The two previous items imply $|\varphi \cdot X| \subseteq |\varphi|$ (recall that $|\varphi|$ is the set of points where φ is *not* locally affine). Again, in general this is not an equality. For example, consider the one-dimensional tropical fan $(\mathbb{R} \times \{0\}) + (\{0\} \times \mathbb{R})$ (whose support is the union of the two coordinate axes in $\mathbb{R}^2$) with the fan function φ who takes the values $(1,0) \mapsto 1, (0,1) \mapsto -1, (-1,0) \mapsto 0, (0,-1) \mapsto 0$. Its Weil divisor $\mathrm{div}(\varphi)$ is empty, even though φ is not affine around 0.

- The Weil divisor construction is independent of the choice of the polyhedral structure $\mathcal{X}$. If we choose a refinement $\mathcal{X}' \triangleright \mathcal{X}$, then for a ridge τ' of $\mathcal{X}'$ we can distinguish two cases. First, if τ' is contained in a ridge τ of $\mathcal{X}$, then there is a one-to-one correspondence between facets of $\mathcal{X}'$ around τ' and facets of $\mathcal{X}$ around τ and the two weight formulas coincide. Secondly, if τ' is not contained in a ridge of $\mathcal{X}$ (but in a facet σ) then the neighbourhood $U(\tau')$ (with respect to $\mathcal{X}'$) is contained in $\mathrm{RelInt}(\sigma)$, where φ is affine. Therefore, by the previous remarks the weight of τ' is 0, as expected.

- Let φ, ψ be two rational functions. It follows from the linearity of the weight formula 1.4 and from $(\varphi + \psi)_\tau = \varphi_\tau + \psi_\tau$ that $\mathrm{div}(\varphi + \psi) = \mathrm{div}(\varphi) + \mathrm{div}(\psi)$, i.e. we constructed a group homomorphism

$$\mathrm{div} : \mathrm{Rat}(X) \to Z_{d-1}(X),$$

whose kernel contains the globally affine forms.

- Let Y be a tropical subcycle of X. Due to remark 1.2.3, the restriction of φ to Y is a well-behaved "non-zero" rational function on Y and we can define

$$\varphi \cdot Y = \mathrm{div}(\varphi|_{|Y|}).$$

The linearity equation $\varphi \cdot (Y + Y') = \varphi \cdot Y + \varphi \cdot Y'$ is a direct consequence of the definition of sums of cycles. Therefore we get the bilinear *intersection product of rational functions on X and subcycles of X*

$$\cdot : \mathrm{Rat}(X) \times Z_*(X) \to Z_{*-1}(X).$$

Moreover, we can also form multiple intersection products $\varphi_1 \cdots \varphi_n \cdot X$. For example, we can restrict a rational function to its own Weil divisor and form $\varphi^2 \cdot X$.

- If F is a tropical fan and φ is a fan function on F, then $\varphi \cdot F$ is again a tropical fan.

Example 1.2.8

The following picture illustrates the constructions of the Weil divisors of the functions $\max\{x, 0\} : \mathbb{R} \to \mathbb{R}$ and $\max\{x, y, 0\} : \mathbb{R}^2 \to \mathbb{R}$.

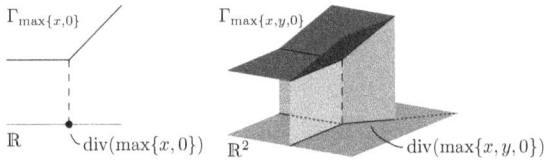

We obtain $\mathrm{div}(\max\{x, 0\}) = \{0\}$ and $\mathrm{div}(\max\{x, y, 0\}) = L_1^2$. More general, one can easily check that

$$\max\{x_1, \ldots, x_r, 0\}^{r-d} \cdot \mathbb{R}^r = L_d^r.$$

The following example of a multiple intersection product will be helpful later on. It deals with the tropical intersection of classical hyperplanes.

Lemma 1.2.9

Let $h_1, \ldots, h_l$ be integer linear functions on V ($l \leq \dim(V) =: r$) and define the rational functions $\varphi_i := \max\{h_i, 0\}$ on V. Let $H : V \to \mathbb{R}^l$ be the linear function with $H(x) = (h_1(x), \ldots, h_l(x))$ and let us assume that H has full rank. Then $\varphi_1 \cdots \varphi_l \cdot V$ is equal to the subspace $\ker(H)$ with weight $\text{ind}(H) := |\mathbb{Z}^l / H(\Lambda)|$.

Proof. Throughout the proof, we give V the fan structure consisting of all cones where each of the h_i is either positive or zero or negative, i.e. $\mathcal{H} = \mathcal{H}_{h_1} \cap \ldots \cap \mathcal{H}_{h_l}$. First, let us assume $l = 1$ (i.e. $H = h_1$). In this case we have to compute the weight of the only ridge in V which is $h_1^\perp = \ker(H)$. This ridge is contained in the two facets corresponding to $h_1 \geq 0$ and $h_1 \leq 0$. Let $v_\geq = -v_\leq$ be corresponding primitive representatives. By definition this implies that for example $v_\geq$ generates the one-dimensional lattice $\Lambda/h_1^\perp$ and therefore $|\mathbb{Z}/h_1(\Lambda)| = h_1(v_\geq)$. Therefore the weight of $h_1^\perp$ in $\varphi_1 \cdot V$ is

$$\omega_{\varphi_1 \cdot V}(h_1^\perp) = \varphi_1(v_\geq) + \varphi_1(v_\leq) = h_1(v_\geq) + 0 = |\mathbb{Z}/h_1(\Lambda)|.$$

Now we use induction on $l > 1$. The induction hypothesis says that $\varphi_2 \cdots \varphi_l \cdot V$ is equal to the subspace $\ker(H')$ with weight $\text{ind}(H')$, where $H' = h_2 \times \ldots \times h_l$. By applying the case $l = 1$ to the vector space $\ker(H') = (\ker(H') \cap \Lambda) \otimes \mathbb{R}$, we obtain that $\varphi_1 \cdots \varphi_l \cdot V$ is equal to the subspace $h_1^\perp \cap \ker(H') = \ker(H)$ with weight $\text{ind}(h_1|_{\ker(H')}) \cdot \text{ind}(H')$. We have to show that this weight coincides with $\text{ind}(H)$. This follows from the exact sequence

$$
\begin{array}{ccccccc}
0 & \to & h_1(\ker(H') \cap \Lambda) & \to & H(\Lambda) & \to & H'(\Lambda) & \to & 0 \\
 & & h_1(x) & \mapsto & (h_1(x), 0) = H(x) & & & & \\
 & & & & H(x) & \mapsto & H'(x) & &
\end{array}
$$

and its induced quotient sequence

$$0 \; \to \; \mathbb{Z}^{l-1}/H'(\Lambda) \; \to \; \mathbb{Z}^l/H(\Lambda) \; \to \; \mathbb{Z}/h_1(\ker(H') \cap \Lambda) \; \to \; 0 \, .$$

$\square$

Remark 1.2.10

In the special case $l = r$ (which can always be achieved by computing locally modulo $\ker(H)$, cf. next subsection) the weight of $\{0\}$ in the intersection product $\varphi_1 \cdots \varphi_r \cdot V$ is $|\mathbb{Z}^r/H(\Lambda)|$, which equals $|\det(M)|$ where M is a matrix representation of H with respect to a lattice basis of Λ and the standard basis of $\mathbb{Z}^r$. Note that it can be extended to the case where H has not full rank, as then the intersection product as well as the determinant $\det(M)$ are zero.

Another reformulation goes as follows. Let $\Lambda_H^{\vee} := \ker(H)^{\perp} \cap \Lambda^{\vee}$ be the saturated sublattice of $\Lambda^{\vee}$ generated by the h_i. Then the equation

$$\mathrm{ind}(H) = [\Lambda_H^{\vee} : \mathbb{Z}h_1 + \ldots + \mathbb{Z}h_l]$$

holds.

1.2.3 Locality, balancing condition and commutativity of $\varphi \cdot X$

As mentioned above, the justification for the Weil divisor definition is still missing: We have to check if $\mathrm{div}(\varphi)$ satisfies the balancing condition. We prove this now, together with another crucial property of our intersection product, commutativity. Before that, it is helpful to analyze the locality of our intersection product more thoroughly.

Let $\mathcal{X}$ be a weighted polyhedral complex and let τ be a cell of $\mathcal{X}$. We would like to make the neighbourhood $U(\tau)$ into a weighted fan as well. We do this by dividing out V_{τ}. Let $q : V \to V/V_{\tau}$ be the corresponding quotient map and denote by $\bar{\sigma}$ the cone spanned by the image of $\sigma - \tau$ under q. Then we define the *star of $\mathcal{X}$ at τ* to be the fan

$$\operatorname{Star}_{\mathcal{X}}(\tau) := \{\bar{\sigma}|\tau < \sigma \in \mathcal{X}\}$$

with weights $\omega(\bar{\sigma}) := \omega(\sigma)$ (note that q preserves the codimension of the cells). Of course, $\mathcal{X}$ is balanced in $U(\tau)$ if and only if $\operatorname{Star}_{\mathcal{X}}(\tau)$ is balanced (i.e. forms a tropical fan denoted by $\operatorname{Star}_X(\tau)$). This star fan is convenient to describe the local structure of X at τ. Note that for any codimension one pair $\sigma' < \sigma$, the class of a primitive representative $q(v_{\sigma/\sigma'})$ modulo V_τ gives a primitive representative $v_{\bar{\sigma}/\bar{\sigma}'}$ of the quotient cells. In particular, if $\sigma' = \tau$ we get $q(v_{\sigma/\tau}) = u_{\sigma/\tau} = u_{\bar{\sigma}/\{0\}}$.

Remark 1.2.11

Let $\mathcal{X}''$ be a refinement of $\mathcal{X}$ and fix a pair of cells $\mathcal{X}' \ni \tau' \subseteq \tau \in \mathcal{X}$ where τ is chosen minimally. Then $\operatorname{Star}_X(\tau)$ equals $\operatorname{Star}_X(\tau')$ divided by $V_\tau/V_{\tau'}$ (which is certainly contained in the space of lineality of $\operatorname{Star}_X(\tau')$).

Let furthermore φ be a rational function on $U(\tau)$. Choose an arbitrary affine form ψ with $\varphi|_\tau = \psi|_\tau$. Then $\varphi - \psi$ induces a rational function φ^τ on $\operatorname{Star}_{\mathcal{X}}(\tau)$ such that $\varphi^\tau|_{\bar{\sigma}}$ is equal to the linear part $(\varphi - \psi)_\sigma \in V_\tau^\perp$ (i.e. such that $(\varphi - \psi)(x) = \varphi^\tau \circ q(x - \tau)$). We call φ^τ a *germ of φ at τ*. Due to the choice of ψ, this function is unique only up to adding a linear form. However, we mentioned in remark 1.2.7 that this indeterminacy does not influence the Weil divisor and the intersection-theoretic behaviour of φ^τ, which is enough for our purposes. The following proposition states the locality of our intersection product in terms of stars and germs.

Proposition 1.2.12 (Locality)

Let X be a balanced polyhedral complex with cells $\tau < \sigma \in X$. Let φ, $\varphi_1, \ldots \varphi_l$ be rational functions on X. Then the following statements are true:

(a) $\mathrm{Star}_{\mathrm{Star}_{\mathcal{X}}(\tau)}(\bar{\sigma}) = \mathrm{Star}_{\mathcal{X}}(\sigma)$

(b) $(\varphi^\tau)^\sigma = \varphi^\sigma$ on $\mathrm{Star}_{\mathcal{X}}(\sigma)$ (up to adding a linear form)

(c) $\mathrm{Star}_{\varphi \cdot \mathcal{X}}(\tau) = \varphi^\tau \cdot \mathrm{Star}_{\mathcal{X}}(\tau)$

(d) $\mathrm{Star}_{\varphi_1 \cdots \varphi_l \cdot \mathcal{X}}(\tau) = \varphi_1^\tau \cdots \varphi_l^\tau \cdot \mathrm{Star}_{\mathcal{X}}(\tau)$

(e) If $l = \dim(\mathcal{X}) - \dim(\tau)$, then

$$\omega_{\varphi_1 \cdots \varphi_l \cdot \mathcal{X}}(\tau) = \omega_{\varphi_1^\tau \cdots \varphi_l^\tau \cdot \mathrm{Star}_{\mathcal{X}}(\tau)}(\{0\}),$$

i.e. we can compute the weight of τ in $\varphi_1 \cdots \varphi_l \cdot \mathcal{X}$ "locally" in $\mathrm{Star}_{\mathcal{X}}(\tau)$.

Proof. (a) and (b) are immediate consequences of the definitions. (d) follows from (c) by induction and (e) is just a special case of (d), namely when $\varphi_1^\tau \cdots \varphi_l^\tau \cdot \mathrm{Star}_{\mathcal{X}}(\tau)$ is zero-dimensional. Hence we are left to show (c).

Let $k := \dim(\mathcal{X}) - \dim(\tau)$ be the codimension of τ in $\mathcal{X}$. The statement is trivial when $k = 0$: Both sides equal the zero cycle $\emptyset$. Assume $k = 1$. In this case, we only have to check

$$\omega_{\varphi \cdot \mathcal{X}}(\tau) = \omega_{\varphi^\tau \cdot \mathrm{Star}_{\mathcal{X}}(\tau)}(\{0\}),$$

(which equals (e) in the case $l = 1$). By adding an affine form ψ we can assume that $\varphi|_\tau \equiv 0$ without changing the intersection product and in particular the weight of τ in $\varphi \cdot \mathcal{X}$. But then we can replace both weights according to their definition and observe that

$$
\begin{aligned}
\omega_{\varphi \cdot \mathcal{X}}(\tau) &= \sum_{\substack{\sigma \in \mathcal{X}^{(\dim(\mathcal{X}))} \\ \tau < \sigma}} \omega(\sigma) \varphi_\sigma(v_{\sigma/\tau}) \\
&= \sum_{\bar{\sigma} \in \mathrm{Star}_{\mathcal{X}}(\tau)^{(1)}} \omega(\bar{\sigma}) \varphi^\tau(u_{\bar{\sigma}/\{0\}}) = \omega_{\varphi^\tau \cdot \mathrm{Star}_{\mathcal{X}}(\tau)}(\{0\})
\end{aligned}
$$

holds true, as $q(v_{\sigma/\tau}) = u_{\bar\sigma/\{0\}} \in V/V_\tau$.

Now let us assume $k > 1$ and let τ' be a ridge in $\mathcal{X}$. Then we can use the previous case as well as (a) and (b) to obtain

$$\omega_{\varphi \cdot \mathcal{X}}(\tau') \overset{k=1}{=} \omega_{\varphi^{\tau'} \cdot \mathrm{Star}_{\mathcal{X}}(\tau')}(\{0\})$$

$$\overset{(a),\,(b)}{=} \omega_{(\varphi^\tau)^{\tau'} \cdot \mathrm{Star}_{\mathrm{Star}_{\mathcal{X}}(\tau)}(\tau')}(\{0\}) \overset{r=1}{=} \omega_{\varphi^\tau \cdot \mathrm{Star}_{\mathcal{X}}(\tau)}(\bar\tau'),$$

which proves the claim. $\qquad\qquad\qquad\qquad\qquad\qquad\qquad\qquad\qquad\quad\Box$

Proposition 1.2.13 (Balancing condition and commutativity)

Let X be a tropical d-cycle in V and $\varphi \in \mathrm{Rat}(X)$ a rational function on X.

(a) Then the Weil divisor $\mathrm{div}(\varphi) = \varphi \cdot C$ is balanced.

(b) Let $\psi \in \mathrm{Rat}(X)$ be another rational function on X. Then it holds
$$\psi \cdot (\varphi \cdot X) = \varphi \cdot (\psi \cdot X).$$

Proof. (a): Choose a polyhedral structure $\mathcal{X}$ of X on whose cones φ is affine. We have to check the balancing condition of $\varphi \cdot \mathcal{X}$ around each cell $\theta \in \mathcal{X}^{(d-2)}$ of codimension 2 in $\mathcal{X}$. By locality of the intersection product this is equivalent to show that $\varphi^\theta \cdot \mathrm{Star}_{\mathcal{X}}(\theta)$ is balanced. In other words, we can restrict the proof to the situation where $\mathcal{X}$ is a two-dimensional fan, $\theta = \{0\}$ and φ is a fan function on $\mathcal{X}$. Moreover, by proposition 1.1.2 we can assume that $\mathcal{X}$ is a unimodular fan.

In this situation, each two-dimensional cone $\sigma \in \mathcal{X}^{(2)}$ is generated by two unique rays $\tau, \tau' \in \mathcal{X}^{(1)}$, i.e. $\sigma = \tau + \tau'$, Moreover, a primitive representative of σ modulo τ is given by the primitive generator $u_{\tau'} := u_{\tau'/\{0\}} \in V$ of the ray τ'. This is because σ is unimodular and therefore $\Lambda_\sigma = \mathbb{Z}u_\tau + \mathbb{Z}u_{\tau'}$. This means that we can rewrite the balancing condition of $\mathcal{X}$ around $\tau \in \mathcal{X}^{(1)}$ only using the vectors generating the rays, namely

$$\sum_{\substack{\tau' \in \mathcal{X}^{(1)} \\ \tau + \tau' \in \mathcal{X}^{(2)}}} \omega(\sigma) u_{\tau'} \in V_\tau$$

$$= \alpha_\tau u_\tau,$$

where α_τ is a coefficient in $\mathbb{R}$ and σ denotes $\tau + \tau'$ in this sum (and the following ones). The weight $\omega_\varphi(\tau)$ of τ in $\operatorname{div}(\varphi)$ can then be computed as

$$\omega_\varphi(\tau) = \left(\sum_{\substack{\tau' \in \mathcal{X}^{(1)} \\ \tau + \tau' \in \mathcal{X}^{(2)}}} \omega(\sigma)\varphi(u_{\tau'}) \right) - \alpha_\tau \varphi(u_\tau).$$

Note that we can omit taking the linear parts of φ due to $\varphi(0) = 0$. Let us now check the balancing condition of $\varphi \cdot \mathcal{X}$ around $\{0\}$ by plugging in these equations. We have to show that

$$\sum_{\tau \in \mathcal{X}^{(1)}} \omega_\varphi(\tau) u_\tau = \sum_{\substack{\tau, \tau' \in \mathcal{X}^{(1)} \\ \tau + \tau' \in \mathcal{X}^{(2)}}} \omega(\sigma)\varphi(u_{\tau'}) u_\tau - \sum_{\tau \in \mathcal{X}^{(1)}} \alpha_\tau \varphi(u_\tau) u_\tau.$$

vanishes. By commuting τ and τ' in the first summand we get

$$\sum_{\tau \in \mathcal{X}^{(1)}} \omega_\varphi(\tau) u_\tau = \sum_{\substack{\tau, \tau' \in \mathcal{X}^{(1)} \\ \tau + \tau' \in \mathcal{X}^{(2)}}} \omega(\sigma)\varphi(u_\tau) u_{\tau'} - \sum_{\tau \in \mathcal{X}^{(1)}} \alpha_\tau \varphi(u_\tau) u_\tau$$

$$= \sum_{\tau \in \mathcal{X}^{(1)}} \varphi(u_\tau) \left(\underbrace{\left(\sum_{\substack{\tau' \in \mathcal{X}^{(1)} \\ \tau + \tau' \in \mathcal{X}^{(2)}}} \omega(\sigma) u_{\tau'} \right) - \alpha_\tau u_\tau}_{= 0 \text{ (balancing condition around } \tau)} \right)$$

$$= 0.$$

This finishes the proof of (a).

(b): We have to check for any $\theta \in \mathcal{X}^{(d-2)}$ that the weights

$$\omega_{\psi \cdot (\varphi \cdot \mathcal{X})}(\theta) = \omega_{\varphi \cdot (\psi \cdot \mathcal{X})}(\theta)$$

coincide. Again, we can use locality and proposition 1.1.2 and restrict to the case where $\mathcal{X}$ is a two-dimensional unimodular fan, $\theta = \{0\}$ and φ is a fan function. With the notations and the same trick as in (a) we get

$$\omega_{\psi \cdot (\varphi \cdot \mathcal{X})}(\{0\}) = \sum_{\substack{\tau, \tau' \in \mathcal{X}^{(1)} \\ \tau + \tau' \in \mathcal{X}^{(2)}}} \omega(\sigma) \varphi(u_{\tau'}) \psi(u_\tau) - \sum_{\tau \in \mathcal{X}^{(1)}} \alpha_\tau \varphi(u_\tau) \psi(u_\tau)$$

$$= \omega_{\varphi \cdot (\psi \cdot \mathcal{X})}(\{0\}),$$

which finishes part (b). $\qquad\qquad\qquad\qquad\qquad\qquad\qquad\qquad\qquad\square$

1.2.4 Relations to toric intersection theory

The given construction of the Weil divisor of a rational function and the proofs of balancing condition and commutativity have the advantage that their kind of reasoning lies completely inside tropical geometry. We only use the combinatorial structure of the objects without referring to any part of classical geometry. The disadvantage of this approach is that the results are still somewhat mysterious and one does not really get a feeling *why* things work. It is therefore helpful to relate our constructions to toric geometry again.

Tropical fans and toric varieties

Let X be a tropical fan of dimension d. The dimension of the ambient vector space $V = \Lambda \otimes \mathbb{R}$ is denoted by r. Let φ be a fan function on X. We choose a fan Ω with the following properties:

- Ω is complete,

- Ω is unimodular,

- X is Ω-directional, i.e. $|X| \subseteq |\Omega^{(d)}|$,

- φ is linear on the cones of Ω (if defined).

Note that such a fan Ω always exists: First, choose a fan structure of X on whose cones φ is linear. Then the construction of lemma 1.1.5 provides a complete fan $\mathcal{H}$ (if $\{0\}$ is not contained, intersect with an arbitrary complete fan) fulfilling all above properties but unimodularity. Thus Ω can be chosen to be a unimodular refinement of $\mathcal{H}$, which exists by proposition 1.1.2.

Let $\mathcal{X} = \{\tau \in \Omega | \tau \subseteq |X|\}$ be the associated fan structure of X. Then the associated toric variety $\mathbf{X}(\mathcal{X})$ is an open subvariety of $\mathbf{X} := \mathbf{X}(\Omega)$. By corollary 1.1.16, X defines a cohomology class $\gamma \in A^{r-d}(\mathbf{X})$. By theorem 1.2.4, φ defines a torus-equivariant Cartier divisor η on $\mathbf{X}(\mathcal{X})$. We choose an arbitrary torus-equivariant extension $\tilde{\eta}$ of η to $\mathbf{X}$, which corresponds to extending φ to Ω. (Such an extension always exists: We assign arbitrary integer values to the primitive generators of all rays not belonging to X and extend the function linearly on all cones, cf. 1.2.2 (e)). Intersecting cycles of $\mathbf{X}$ with $\tilde{\eta}$ defines a cohomology class in $A^1(\mathbf{X})$, which we denote by $[\tilde{\eta}]$ (in fact, as $\mathbf{X}$ is a complete toric variety, we have $\mathrm{Pic}(\mathbf{X}) \cong A^1(\mathbf{X})$, cf. [Bri89] or [FS94, corollary 2.4]). Therefore we can form the cup-product of cohomology classes

$$\kappa := \gamma \cup [\tilde{\eta}] \in A^{r-(d-1)}(\mathbf{X}).$$

Now, we can go back: κ defines a $(d-1)$-dimensional tropical fan Y_κ contained in $\Omega^{(d)}$. And we can prove:

Theorem 1.2.14

The classical toric and the tropical construction provide the same result, i.e.

$$Y_\kappa = \varphi \cdot X.$$

In particular, Y_κ and κ do not depend on the choice of an extension $\tilde{\eta}$.

Proof. Recall from subsection 1.1.4 that the correspondence between the Ω-directional tropical fan $Y := Y_\kappa$ and the cohomology class κ of $\mathbf{X}$ is given via

$$\omega_Y(\tau) = \deg(\kappa \cap [V(\tau)]), \tag{1.5}$$

where $\tau \in \Omega^{(d-1)}$, $V(\tau)$ denotes the closure of the $r - (d-1)$-dimensional orbit corresponding to τ and $\cap$ denotes the intersection pairing of cohomology and homology, providing a zero-dimensional cycle in this case. So let us first compute $[\tilde{\eta}] \cap [V(\tau)]$. Note that $V(\tau)$ is also a toric variety, namely $V(\tau) = \mathbf{X}(\mathrm{Star}_\Omega(\tau))$. Let $\lambda \subset \Lambda$ be a linear form and let x^λ be the associated rational function on $\mathbf{X}$. Its (classical) Weil divisor is given by the formula (cf. [Fu93, section 3.3])

$$\mathrm{div}(x^\lambda) = \sum_{\varrho \in \Omega^{(1)}} \lambda(u_{\varrho/\{0\}})[V(\varrho)], \tag{1.6}$$

from which we deduce as a first implication that x^λ can be restricted to a non-zero rational function on $V(\tau)$ if and only if λ vanishes on τ, i.e. $\lambda \in \tau^\perp$. This means that, in order to pull back $[\tilde{\eta}]$ to $V(\tau)$, we have to divide $\tilde{\eta}$ by a rational function x^λ with $\lambda|_\tau = \varphi|_\tau$, such that all local equations of $\tilde{\eta}/x^\lambda$ can be restricted to $V(\tau)$. In other words, the pull back of $[\tilde{\eta}]$ to $V(\tau)$ is represented by a germ $\tilde{\varphi}^\tau$ on $\mathrm{Star}_\Omega(\tau)$. Applying formula 1.6 again to (the local equations of) $[\tilde{\eta}]|_{V(\tau)}$, we get

$$[\tilde{\eta}] \cap [V(\tau)] = \sum_{\substack{\sigma \in \Omega^{(d)} \\ \tau < \sigma}} \tilde{\varphi}^\tau(u_{\sigma/\tau})[V(\sigma)] \in A_{r-d}(\mathbf{X}).$$

On the other hand, we know how to intersect the cohomology class γ with the orbit closures $V(\sigma)$. Namely, formula 1.5 applied to γ provides

$$\deg(\gamma \cap [V(\sigma)]) = \omega_{\mathcal{X}}(\sigma) = \omega_{\mathrm{Star}_\Omega(\tau)}(\bar\sigma),$$

where $\omega_{\mathcal{X}}(\sigma) = 0$ if $\sigma \notin \mathcal{X}$. Putting things together, we get

$$
\begin{aligned}
\deg(\kappa \cap [V(\tau)]) &= \deg(\gamma \cap ([\tilde\eta] \cap [V(\tau)])) \\
&= \sum_{\substack{\sigma \in \Omega^{(d)} \\ \tau < \sigma}} \tilde\varphi^\tau(u_{\sigma/\tau}) \omega_{\mathrm{Star}_\Omega(\tau)}(\bar\sigma) \\
(\text{by omitting } \sigma \notin \mathcal{X}) \quad &= \sum_{\substack{\sigma \in \mathcal{X}^{(d)} \\ \tau < \sigma}} \tilde\varphi^\tau(u_{\sigma/\tau}) \omega_{\mathrm{Star}_{\mathcal{X}}(\tau)}(\bar\sigma) \\
(\text{by locality}) \quad &= \omega_Y(\tau).
\end{aligned}
$$

This proves the claim. $\qquad\qquad\qquad\qquad\qquad\qquad\qquad\qquad\qquad$ □

Remark 1.2.15

Let us make two remarks here.

- The independence of the chosen extension $\tilde\eta$ is also obvious from the classical side. Due to the properties of γ, the relevant computations with $\tilde\eta$ take place at orbit closures $V(\sigma), \sigma \in \mathcal{X}$. As these orbit closures have non-empty intersection with the open set $\mathbf{X}(\mathcal{X}) \subseteq \mathbf{X}$, it suffices to consider $\tilde\eta|_{\mathbf{X}(\mathcal{X})} = \eta$.

- As the balancing condition of Y is guaranteed by the classical theory (cf. subsection 1.1.4), we could replace the proof that $\varphi \cdot X$ is balanced by the above argument. However, it is important to keep in mind that our tropical intersection product does not involve the choice of a specific toric variety. Instead, the above argument works for any appropriate fan Ω.

Another consequence of the above connections can be formulated as follows. Let $V = \Lambda \otimes \mathbb{R}$ be an r-dimensional vector space and let $P_1, \dots, P_r$

be polytopes in $V^\vee$ obtained as the convex hull of a finite number of lattice vectors. For every face τ of P_i, we can form the cone of elements v in V such that the face where v is minimal on P_i contains τ. The cones for all $\tau < P_i$ form the dual fan $\Omega(P_i)$. Assume that Ω is a refinement of all these dual fans $\Omega(P_i)$. Then the P_i define Cartier divisors γ_i on $\mathbf{X}$ with associated rational functions

$$\varphi_i(v) = -\min_{\lambda \in P_i \cap \Lambda^\vee} \{\lambda(v)\}$$

(cf. [Fu93, section]; recall that due to a different notation we get a minus sign here). It is discussed in [Fu93, section 5.4] that the degree of the intersection of these Cartier divisors can be computed as the mixed volume of the polytopes P_i

$$\deg(\gamma_1 \cup \ldots \cup \gamma_r) = \mathrm{MV}(P_1, \ldots, P_r).$$

The *mixed volume* is given by the formula

$$\mathrm{MV}(P_1, \ldots, P_r) := \mathrm{Vol}(P_1 + \ldots + P_r)$$
$$- \sum_{i=1}^{r} \mathrm{Vol}(P_1 + \ldots + \widehat{P}_i + \ldots + P_r)$$
$$+ \sum_{i<j} \mathrm{Vol}(P_1 + \ldots + \widehat{P}_i + \ldots + \widehat{P}_j + \ldots + P_r)$$
$$- \ldots$$
$$+ (-1)^{r-1} \sum_{i=1}^{r} \mathrm{Vol}(P_i),$$

where the volume is normalized by the lattice cube. Hence we can state the following corollary.

Corollary 1.2.16

The mixed volume of the polytopes $P_1, \ldots, P_r$ can be computed tropically by

$$\mathrm{MV}(P_1, \ldots, P_r) = \deg(\varphi_1 \cdots \varphi_r \cdot V).$$

Remark 1.2.17

For the sake of completeness, let us also mention the tropical description of the isomorphism $\mathrm{Pic}(\mathbf{X}) \cong A_{n-1}(\mathbf{X})$ by taking Weil divisors (if $\mathbf{X}$ is a smooth toric variety given by the complete unimodular fan Ω). This is the same as considering the isomorphism

$$\mathrm{Div}_T(\mathbf{X}) \cong Z^T_{n-1}(\mathbf{X})$$
$$\gamma \mapsto \mathrm{div}(\gamma),$$

and then dividing out rational equivalence

$$x^\lambda \mapsto \mathrm{div}(x^\lambda)$$

on both sides. Here, $\mathrm{Div}_T(\mathbf{X})$ and $Z^T_{n-1}(\mathbf{X})$ denote the group of torus-equivariant Cartier and Weil divisors. Hence $Z^T_{n-1}(\mathbf{X})$ can be identified with $\mathbb{Z}^{\Omega^{(1)}}$ and it is easy to check with formula 1.6 on page 44 that the "tropical" version of this isomorphism is

$$\mathrm{Rat}(\Omega) \cong \mathbb{Z}^{\Omega^{(1)}}$$
$$\varphi \mapsto (\varphi(u_{\varrho/\{0\}}))_\varrho,$$
$$\sum_\varrho a_\varrho \varphi_\varrho \leftarrow\!\shortmid (a_\varrho)_\varrho,$$

(cf. example 1.2.2 (e) for φ_ϱ), i.e. one way is evaluating a rational function at the primitive generators of the rays, and the way back is defining a rational function by linear extension on the cones from its values on the u_ϱ. Dividing out rational equivalence corresponds to dividing out the linear forms $\Lambda^{\vee}$ and their image on the right hand side.

Tropical cycles and toric degenerations

The above construction for fans and fan function can also be generalized to polyhedral complexes and rational functions. This refers to so-called toric degenerations, which will appear also in section 1.6. As I could not find a reference brief and simple enough for what is needed here, let me give a small account.

Let Σ be a polyhedral subdivision of V (i.e. a polyhedral complex whose support equals V) such that all cells contain vertices (i.e. have trivial space of lineality). We now perform a construction similar to toric varieties with the only difference that we replace notions of linear geometry by those of affine geometry.

For any cell $\tau \in \Sigma$, we denote by

$$\tau^{\vee} := \{\alpha \text{ integer affine form} \mid \alpha(x) \geq 0 \text{ for all } x \in \tau\}$$

the monoid of integer affine forms which take no negative values on τ and by

$$\tau^{\perp} := \{\alpha \text{ integer affine form} \mid \alpha(x) = 0 \text{ for all } x \in \tau\}$$

the space of integer affine forms vanishing on τ. For all $\alpha \in \tau^{\vee}$, we define $\min(\alpha, \tau) := \min_{x \in \tau}\{\alpha(x)\}$ and denote the face of τ where this minimum is attained by $\text{face}(\alpha, \tau)$. The associated algebra to $\tau^{\vee}$ (over some algebraically closed field κ) is denoted by $\kappa[\tau^{\vee}]$, its elements are polynomials of the form $f = \sum_{\alpha \in S} a_{\alpha} x^{\alpha}$, where $S \subseteq \tau^{\vee}$ is a finite subset and a_{α} are coefficients in κ. The algebra we are really interested in is the quotient $A_{\tau} := \kappa[\tau^{\vee}]/I_{\tau}$ by the ideal

$$I_{\tau} := \langle x^{\alpha} \mid \min(\alpha, \tau) > 0 \rangle.$$

Note that all such x^α are not invertible in $\kappa[\tau^\vee]$ as $-\alpha$ necessarily attains negative values on τ. Note also that the remaining monomials might be zero-divisors now: A product $x^\alpha \cdot x^{\alpha'}$ with $\min(\alpha, \tau) = \min(\alpha', \tau) = 0$ is zero in A_τ if and only if $\min(\alpha + \alpha', \tau) > 0$, if and only if $\mathrm{face}(\alpha, \tau) \cap \mathrm{face}(\alpha', \tau) = \emptyset$. In particular, A_τ is an integral domain if and only if τ contains precisely one vertex. Otherwise, the associated scheme $U_\tau = \mathrm{Spec}(A_\tau)$ is reducible. Now we glue together these open patches as in the case of toric varieties. For a pair $\tau < \sigma$ and an affine form $\alpha \in \sigma^\vee$ with $\mathrm{face}(\alpha, \sigma) = \tau$, it is straightforward to verify $(A_\sigma)_{(x^\alpha)} = A_\tau$ (obviously $\alpha \in \tau^\perp$), i.e. U_τ is an open subset of U_σ in a canonical way, and we can glue all $U_\sigma, U_{\sigma'}$ along the the open subset $U_{\sigma \cap \sigma'}$.

The resulting variety $\mathbf{X} := \mathbf{X}(\Sigma)$ is a union of the toric varieties $V(\tau) := \mathbf{X}(\mathrm{Star}_\Sigma(\tau))$ for all cells $\tau \in \Sigma$ (follows easily from identifying $\tau^\perp$ with $\Lambda_\tau^\perp = (\Lambda/\Lambda_\tau)^\vee$ and then applying the arguments from the usual toric case). The irreducible components of $\mathbf{X}$ are the toric varieties $V(\nu)$ for every vertex $\nu \in \Sigma$. The algebraic torus $\mathrm{Hom}(\Lambda^\vee, \kappa^*)$ acts on $\mathbf{X}$ by $x^\alpha \mapsto x^\alpha \cdot t^{\mathrm{LP}(\alpha)}$, where $\mathrm{LP}(\alpha)$ denotes the linear part of α. As for toric varieties, the orbits, which are of the form $\mathrm{Spec}(\kappa[\tau^\perp])$, are in dimension reversing bijection with the cells $\tau \in \Sigma$, and $V(\tau)$ is the closure of the corresponding orbit. For example, the polyhedral complex displayed below provides a variety $\mathbf{X}$ which consists of two copies of $\mathbb{P}_\kappa^2$ (corresponding to the vertices) glued together along a $\mathbb{P}_\kappa^1$ (corresponding to the bounded edge τ).

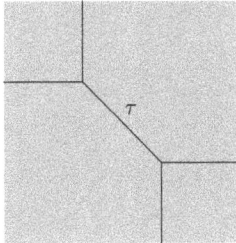

We call $\mathbf{X}$ a toric degeneration (which will be justified in section 1.6). Now, as in the usual toric case, we find the following connection between classical and tropical objects.

As the irreducible components are usual toric varieties, it follows that the homology groups $A_*(\mathbf{X})$ are generated by the orbit closures $V(\tau)$ with relations generated by the divisors

$$\mathrm{div}(x^\lambda) = \sum_{\sigma > \tau} \lambda(u_{\sigma/\tau})[V(\sigma)],$$

where τ is a fixed cell, λ is an integer linear form on V/V_τ and the sum runs through cells σ with $\dim(\sigma) = \dim(\tau) + 1$.

As $\mathbf{X}$ is a "complete scheme on which a connected solvable linear algebraic group acts with finitely many orbits", we know by [FMSS95, theorem 3] that the Kronecker duality homomorphism

$$\mathcal{D}_\mathbf{X} : \Lambda^*(\mathbf{X}) \to \mathrm{Hom}(A_*(\mathbf{X}), \mathbb{Z})$$
$$\gamma \mapsto \deg(\gamma \cap \,.\,)$$

is an isomorphism. Hence we get immediately the following statement.

Corollary 1.2.18
Let Σ be a polyhedral subdivision of V and let $\mathbf{X} = \mathbf{X}(\Sigma)$ be the associated toric degeneration. Furthermore, we denote by $Z_d(\Sigma)$ the group of d-dimensional tropical cycles contained in $|\Sigma^{(d)}|$. Then this group isomorphic to the cohomology group of codimension $r - d$ of $\mathbf{X}$ by

$$A^{r-d}(\mathbf{X}) \to Z_d(\Sigma)$$
$$\gamma \mapsto \left[\omega_\gamma(\sigma) = \deg(\gamma \cap [V(\sigma)])\right].$$

Now let φ be a rational function on V affine on the cells of Σ. It defines a Cartier divisor η on $\mathbf{X}$ as follows. For every cell $\tau \in \Sigma$ and vertex $\nu \in \tau$, a local equation of η around $V(\tau)$ on the irreducible component $V(\nu)$ is given by $x^{\alpha - \alpha(\nu)} \in A_\nu$, where α is an integer affine form such that

$\alpha|_\tau = \varphi$. The condition that the different pieces of φ fit together ensures that this provides a well-defined Cartier divisor. It is easy to check that, as before, the restriction of η to an orbit closure $V(\tau)$ is represented by a germ φ^τ on $\mathrm{Star}_\Sigma(\tau)$. Hence, together with theorem 1.2.14 for fans and toric varieties, we get the following statement:

Corollary 1.2.19
Let X be a tropical cycle and let φ be a rational function on X. For every polyhedral subdivision Σ with $X \in Z_d(\Sigma)$ and for every extension of φ to a function on V which is affine on the cells of Σ, let $\gamma \in A^{r-d}(\mathbf{X})$ and $[\eta] \in A^1(\mathbf{X})$ be the associated cohomology classes. Let Y be the tropical cycle associated to $\gamma \cup [\eta] \in A^{r-d+1}(\mathbf{X})$. Then the equation

$$Y = \varphi \cdot X$$

holds, i.e. cup-product and intersection product are equivalent in this case.

1.2.5 Cartier divisors

The procedure of assigning a Weil divisor to a rational function can be easily extended to Cartier divisors. Thereby, (restrictions of) affine forms play the role of regular invertible functions.

Definition 1.2.20 (Cartier divisors)
Let X be a tropical d-cycle. A rational function $\varphi \in \mathrm{Rat}(U)$ on a polyhedral open set U of $|X|$ is called *locally affine* if it is locally (on every neighbourhood $U(\tau)$ for every polyhedral structure, see remark 1.3.2) the restriction of an affine form on V.

A *representative of a Cartier divisor on X* is a finite set of tuples $\{(U_1, \varphi_1), \ldots, (U_l, \varphi_l)\}$, where $\{U_i\}$ is a polyhedral open covering of $|X|$ and $\varphi_i \in \mathrm{Rat}(U_i)$ are rational functions on U_i that only differ in locally affine functions on the overlaps (i.e. for all $i \neq j$, the function

$\varphi_i|_{U_i \cap U_j} - \varphi_j|_{U_i \cap U_j}$ is locally affine).

We define the *sum* of two representatives by $\{(U_i, \varphi_i)\} + \{(V_j, \psi_j)\} = \{(U_i \cap V_j, \varphi_i + \psi_j)\}$, which obviously fulfills the condition on the overlaps again.

We call two representatives $\{(U_i, \varphi_i)\}, \{(V_j, \psi_j)\}$ *equivalent* if $\varphi_i - \psi_j$ is locally affine (where defined) for all i, j, i.e. $\{(U_i, \varphi_i)\} - \{(V_j, \psi_j)\} = \{(W_k, \gamma_k)\}$ with γ_k locally affine functions. Obviously, "+" induces a group structure on the set of equivalence classes of representatives with the neutral element $\{(|X|, c_0)\}$, where c_0 is the constant zero function. This group is denoted by $\mathrm{Div}(X)$ and its elements are called *Cartier divisors on* X.

Remark 1.2.21
Note that, as in the case of rational functions, we can restrict a Cartier divisor on X to an arbitrary subcycle Y by putting $[\{(U_i, \varphi_i)\}]|_Y := [\{(U_i \cap |Y|, \varphi_i|_{U_i \cap |Y|})\}] \in \mathrm{Div}(Y)$ (which is well-defined because the restriction of an affine function is again affine).

Definition 1.2.22 (Weil divisors of Cartier divisors)
Let X be a tropical d-cycle and $\varphi = [\{(U_i, \varphi_i)\}] \in \mathrm{Div}(X)$ a Cartier divisor on X. Let furthermore $\mathcal{X}$ be a polyhedral structure of X compatible with all U_i and such that all φ_i are affine on the cells (where defined). Then we define

$$\varphi \cdot \mathcal{X} := \mathcal{X} \setminus \mathcal{X}^{(d)}$$

with weight function

$$\omega_\varphi(\tau) := \omega_{\varphi_i}(\tau),$$

where i is chosen such that $U(\tau) \subseteq U_i$ and ω_{φ_i} is given by the weight formula 1.4 for Weil divisors of rational functions (see definition 1.2.6). If we choose a different φ_j with $U(\tau) \subseteq U_j$, the difference $\varphi_i - \varphi_j$ on $U(\tau)$ is affine, so the weight of τ does not depend on this choice nor on the choice of a representative of φ.

The tropical cycle associated to $\varphi \cdot \mathcal{X}$ is called the *Weil divisor of φ* and denoted by $\operatorname{div}(\varphi)$ or $\varphi \cdot X$. Analogous to remark 1.2.7 one checks that this definition does not depend on the chosen polyhedral structure.

Example 1.2.23

Consider the tropical curve X indicated in the picture with vertices $(\pm 1/2, \pm 1/2)$ (and given by the intersection product $\max\{1/2, x, y, -x, -y\} \cdot \mathbb{R}^2$, for example).

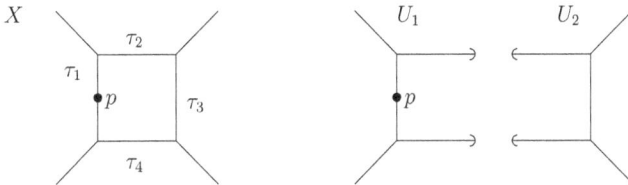

Now we define a Cartier divisor φ on X by the local functions $\varphi_1 = \max\{y, 0\}$ on U_1 and $\varphi_2 \equiv 0$ on U_2 (see picture). Note that that p is the only point where the local functions are not locally affine, and we get $\operatorname{div}(\varphi) = 1 \cdot p$. On the other hand, it is easy to see that there does not exist a rational function ψ on X such that $\operatorname{div}(\psi) = p$: This requires that ψ does not change its slope in the relative interior of τ_2, τ_3 and τ_4 and therefore the values of ψ on the respective boundary vertices can only differ by integers (as ψ must have integer slope). But at p, the function has to break, with a change of slopes by one, so the values at the boundary vertices of τ_1 must differ by an integer plus a half, which is a contradiction.

Remark 1.2.24 • Let τ be a cell in $\mathcal{X}$. As in the case of rational functions, a Cartier divisor φ induces a *germ φ^τ of φ at τ*, a fan function on $\operatorname{Star}_\mathcal{X}(\tau)$ unique up to adding linear functions (i.e. well-defined as Cartier divisor). In fact, φ^τ can be chosen to be the germ of any local function φ_i of φ at τ.

- The Weil divisor of a Cartier divisor can also be computed locally, that is, for any $\tau \in \mathcal{X}$ the equality

$$\mathrm{Star}_{\varphi \cdot \mathcal{X}}(\tau) = \varphi^\tau \cdot \mathrm{Star}_{\mathcal{X}}(\tau)$$

  follows readily from the respective statement for any local function φ_i of φ at τ

- $\varphi \cdot \mathcal{X}$ is balanced and therefore $\varphi \cdot X$ is a well-defined tropical cycle. This also follows directly from the respective statement for rational functions and the previous item.

- The definition makes sure that $|\operatorname{div}(\varphi)| \subseteq |\mathcal{X} \setminus \mathcal{X}^{(d)}|$ holds. In general, equality does not hold.

- Let Y be a subcycle of X. With the definition $\psi \cdot Y = \operatorname{div}(\varphi|_Y)$, we get the bilinear *intersection product of Cartier divisors on X and subcycles of X*

$$\cdot : \operatorname{Div}(X) \times Z_*(X) \to Z_{*-1}(X).$$

  Moreover, we can also form multiple intersection products $\varphi_1 \cdots \varphi_n \cdot X$. These multiple intersection products are commutative, i.e.

$$\psi \cdot (\varphi \cdot X) = \varphi \cdot (\psi \cdot X).$$

  Again, this follows directly from locality and the rational function case.

1.2.6 Convexity, Positivity and Irreducibility

As mentioned before (cf. remark 1.2.7), the equation $|\varphi \cdot X| \subseteq |\varphi|$ is in general not an equality. We will now consider a class of rational functions and cycles where equality holds.

A non-zero cycle X is called *positive*, denoted $X > 0$, if all weights are non-negative (i.e by taking the non-zero part $\mathrm{NZ}(\mathcal{X})$ of a polyhedral structure $\mathcal{X}$, we can get a structure with only positive weights). A rational function φ on X is called *convex* if it is locally the restriction of a convex function on V. The typical example of convex functions are tropical polynomials (min produces concave functions, but the minus sign makes them convex). If Y is a subcycle of X, then $\varphi|_{|Z|}$ is also convex on Z. A Cartier divisor φ is called *convex* if all its local functions are convex. Combining positivity and convexity we get the following result.

Lemma 1.2.25
Let X be a positive tropical cycle and let φ be a convex Cartier divisor on X. Then

(a) $\varphi \cdot X$ is positive and

(b) $|\varphi| = |\varphi \cdot X|$.

Proof. Let $\mathcal{X}$ be a polyhedral structure of X with only positive weights on whose cells φ is affine. As the weights of $\varphi \cdot \mathcal{X}$ can be computed locally, and as convexity and linearity are passed on to the germs φ^τ of φ, we can restrict to the case where $\mathcal{X} = \{\{0\}, \rho_1, \ldots, \rho_r\}$ is a one-dimensional fan with positive weights $\omega(\rho_i) > 0$ and φ is a fan function on $\mathcal{X}$. The statements of the lemma translate to

(a) φ convex $\Rightarrow \omega_{\varphi \cdot \mathcal{X}}(\{0\}) \geq 0$,

(b) φ convex, $\omega_{\varphi \cdot \mathcal{X}}(\{0\}) = 0 \Rightarrow \varphi$ linear.

We use the following criteria for linearity and convexity. Let φ be a fan function on $\mathcal{X}$ and let us abbreviate the primitive vector of the ray ρ_i by v_i. Then

i) φ is linear if and only if for all $\lambda_1, \ldots, \lambda_r \in \mathbb{R}$ with $\sum_i \lambda_i v_i = 0$ the equation

$$\sum_i \lambda_i \varphi(v_i) = 0$$

holds,

ii) φ is convex if and only if for all positive $\lambda_1, \ldots, \lambda_r \geq 0$ with $\sum_i \lambda_i v_i = 0$ the equation

$$\sum_i \lambda_i \varphi(v_i) \geq 0$$

holds.

Now let φ be convex. We can apply criterion ii) to the coefficients $\omega(\rho_i)$, which are positive and satisfy $\sum_i \omega(\rho_i) v_i = 0$. This provides

$$\omega_{\varphi \cdot \mathcal{X}}(\{0\}) = \sum_i \omega(\rho_i) \varphi(v_i) \geq 0,$$

which proves (a).

For (b), let us assume that $\sum_i \omega(\rho_i) \varphi(v_i) = 0$, but φ is not linear. Then by i) there exist $\lambda_1, \ldots, \lambda_r$ with $\sum_i \lambda_i v_i = 0$ but $\sum_i \lambda_i \varphi(v_i) \neq 0$. W.l.o.g. we can assume $\sum_i \lambda_i \varphi(v_i) < 0$ (otherwise we replace λ_i by $-\lambda_i$). For large enough $C \in \mathbb{R}$ the coefficients $\lambda_i' := \lambda_i + C\omega(\rho_i)$ are all positive and still satisfy $\sum_i \lambda_i' v_i = 0$ and $\sum_i \lambda_i' \varphi(v_i) < 0$, which contradicts ii). Therefore φ is linear, which proves (b). $\qquad\square$

Remark 1.2.26 (Irreducible cycles)

Let us mention the toric analogues. Let X be a positive tropical fan and let Ω be an complete unimodular fan such that X is Ω-directional. Then the cohomology class on $\mathbf{X}$ induced by X is positive in the sense that it provides non-negative intersection numbers when intersected with effective cycles. Moreover, let φ be a convex fan function on X linear on the cones of Ω (if defined) and assume there exists an extension of φ to a convex function of Ω. This corresponds to a Cartier divisor/line bundle which is generated by its sections (cf. [Fu93, section 3.4]). Then our lemma states (in the tropical language) that these positivity conditions are preserved when taking the cup-product.

Another class of cycles for which we can say more about the relation $|\varphi \cdot X| \subseteq |\varphi|$ fulfills an local irreducibility condition.

Definition 1.2.27
A tropical cycle X (of dimension d) is called *irreducible* if $Z_d(X) = \mathbb{Z}X$, i.e. if any subcycle of the same dimension d is an integer multiple of X.

A tropical cycle X is called *locally irreducible* if for a chosen polyhedral structure $\mathcal{X}$ the greatest common divisor of all weights is 1 and for all ridges $\tau \in \mathcal{X}$ the star $\mathrm{Star}_X(\tau)$ is is either irreducible or the multiple of a vector space cycle. As a refinement $\mathcal{X}'$ still satisfies this property by remark 1.2.11, the definition is independent of the chosen polyhedral structure.

Example 1.2.28 (a) A tropical cycle $X = a \cdot V$ which is the multiple of a vector space cycle is irreducible if and only if $a = \pm 1$. Because of the greatest common divisor condition the same is true with regard to being locally irreducible.

(b) The tropical cycle $X = \max\{y, -y, x - 1, -x - 1\} \cdot \mathbb{R}^2$ is locally irreducible, even though it contains an edge with weight 2 and therefore the star around (for example) 0 is twice a vector space cycle.

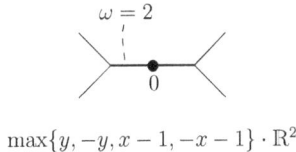

$$\max\{y, -y, x - 1, -x - 1\} \cdot \mathbb{R}^2$$

(c) A one-dimensional tropical fan F is irreducible if and only if the greatest common divisor of all weights is 1 and the number of rays equals $\dim(V_F) + 1$, where V_F denotes the vector space spanned by F. In this case, the balancing condition of F is the unique primitive linear relation which the primitive generators satisfy. In particular, the tropically linear cycles L_1^r from example 1.1.11 are irreducible.

(d) All tropically linear cycles L_d^r are locally irreducible as the weights of L_d^r are all 1 and the stars of all ridges are equivalent to $\mathcal{L}_1^{r-d+1}$. It follows that the cycles L_d^r are also irreducible (cf. next lemma).

An irreducible cycle is always connected in codimension 1. Moreover, this property together with local irreducibility implies irreducibility.

Lemma 1.2.29

Let X be a d-dimensional tropical cycle. If X is locally irreducible and connected in codimension one (i.e. all facets of an (arbitrary) polyhedral structure are connected via ridges), then X is also irreducible.

Proof. Let Y be a d-dimensional subcycle of X. By lemma 1.1.9 we can assume that both are defined by different weight functions ω_X and ω_Y on the same polyhedral complex $\mathcal{X}$. As X is locally irreducible, for every ridge τ we find a coefficient α_τ such that $\operatorname{Star}_Y(\tau) = \alpha_\tau \cdot \operatorname{Star}_X(\tau)$. In particular, for every facet σ containing τ the weights satisfy $\omega_Y(\sigma) = \alpha_\tau \cdot \omega_X(\sigma)$. As the facets are all connected via ridges, all the α_τ have to be equal (say to α) and therefore $\omega_Y = \alpha \cdot \omega_X$ holds, which proves the claim. $\qquad\square$

Remark 1.2.30

It is a deficiency of tropical geometry that a reducible tropical cycle does not admit a unique or canonical decomposition into irreducible components. The following picture shows an example.

Let us now prove the important property of local irreducibility.

Lemma 1.2.31

Let X be a locally irreducible tropical cycle and let φ be a Cartier divisor on X. Then $|\varphi \cdot X| = |\varphi|$ holds.

Proof. By locality of the intersection product, it remains to show: Let $\mathcal{X}$ be a one-dimensional balanced irreducible fan and let φ be a fan function, then $\varphi \cdot X = \emptyset$ implies that φ is linear. In this case, as mentioned in example 1.2.28 (c), irreducibility just means that

$$\sum_{\varrho \in \mathcal{X}^{(1)}} \omega_{\mathcal{X}}(\varrho) u_{\varrho/\{0\}} = 0$$

is the unique primitive linear relation which the primitive generators satisfy. But this precisely means that for given values $\varphi(u_{\varrho/\{0\}})$ such that

$$\sum_{\varrho \in \mathcal{X}^{(1)}} \omega_{\mathcal{X}}(\varrho) \varphi(u_{\varrho/\{0\}}) = 0$$

holds, there exists an integer linear extension attaining these values. This proves the claim. $\qquad\square$

1.3 Morphisms and the projection formula

In this section, our first step is to define morphisms. Of course, a morphism should define functors, namely the push forward of cycles and the pull back of rational functions/Cartier divisors. We show that these functors exist and satisfy expected properties such as the projection formula, which connects the functors to the intersection product.

1.3.1 Morphisms

Let $V = \Lambda \otimes \mathbb{R}$ and $V' = \Lambda' \otimes \mathbb{R}$ be two real vector spaces with underlying lattices Λ and Λ'. A map $f : S \to V'$ from a subset $S \subseteq V$ to V' is called an *integer affine map* if it is the restriction of a map $x \mapsto \Psi(x) + v$, where $\Psi : \Lambda \to \Lambda'$ is an integer linear map and $v \in V'$ is an arbitrary translation vector. Such functions are the local patterns of tropical morphisms.

Definition 1.3.1 (Tropical morphisms)
Let $X \in Z_*(V)$ and $Y \in Z_*(V')$ be tropical cycles. A *tropical morphism* $f : X \to Y$ is a map from $|X|$ to $|Y|$ which is locally integer affine.

A tropical morphism $f : X \to Y$ is an *isomorphism* (and X and Y are *isomorphic*) if there exists a tropical morphism $g : Y \to X$ such that $f \circ g = \mathrm{id}_Y$ and $g \circ f = \mathrm{id}_X$ and if the weights of identified facets of suitable polyhedral structures on X and Y coincide.

Remark 1.3.2 • Obviously, a tropical morphism is continuous. Moreover, for every polyhedral structure $\mathcal{X}$ of X and for every cell $\tau \in \mathcal{X}$, the local function $f|_{U(\tau)}$ is integer affine. Indeed, let $W = \langle U(\tau) \rangle$ be the locally spanned affine space, then the map

$$\mathrm{RelInt}(\tau) \to \mathrm{AffMaps}(W, V')$$

$$x \mapsto \text{local representation of } f$$

is locally constant, i.e. constant. Therefore the restriction of f to a small neighbourhood of a point $x \in \mathrm{RelInt}(\tau)$ fixes f on $U(\tau)$. We denote the *linear part of f at τ* by $f_\tau : \Lambda_{U(\tau)} \to \Lambda'$.

In particular, for every cell $\tau \in \mathcal{X}$ the restriction $f|_\tau$ is integer affine and the image $f(\tau)$ is a polyhedron in V'.

- If the target cycle $Y = V' = \mathbb{Z} \otimes \mathbb{R}$ is the set of real numbers, then we can consider f as a rational function. However, as it is locally affine, its Weil divisor $\mathrm{div}(f)$ is $\emptyset$. We can do something else with such a function. The set

$$\mathcal{H}_f := \Big\{ \{x \in |X| : f(x) \geq 0\},$$
$$\{x \in |X| : f(x) = 0\}, \{x \in |X| : f(x) \leq 0\} \Big\}$$

is not a polyhedral complex, but the intersection

$$\mathcal{X} \cap \mathcal{H}_f = \Big\{ \{x \in \tau : f(x) \geq 0\},$$
$$\{x \in \tau : f(x) = 0\}, \{x \in \tau : f(x) \leq 0\} \text{ for all } \tau \in \mathcal{X} \Big\}$$

with any polyhedral structure $\mathcal{X}$ of X is one. This will be helpful in the following lemma.

Example 1.3.3

Let us give two very simple examples. We take $X = L_1^2$ in $\mathbb{R}^2$ and $Y = \mathbb{R}$ and define the two morphisms $f_1 : (x, y) \mapsto x + y$ and $f_2 : (x, y) \mapsto x$.

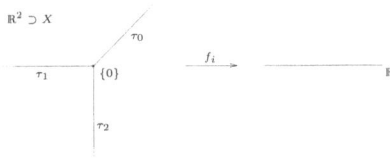

Note that f_1 maps both rays τ_1 and τ_2 onto $\mathbb{R}_{\leq 0}$, whereas f_2 contracts the whole ray τ_2 to the point 0 but is injective otherwise.

The following lemma clarifies the combinatorial situation of tropical morphisms. The proof is similar to [GKM07, construction 2.24].

Lemma 1.3.4

Let $f : X \to Y$ be a tropical morphism. Then there exist polyhedral structures $\mathcal{X}$ of X and $\mathcal{Y}$ of Y such that the polyhedron $f(\tau)$ is contained in $\mathcal{Y}$ for any $\tau \in \mathcal{X}$, i.e.

$$\{f(\tau) | \tau \in \mathcal{X}\} \subseteq \mathcal{Y}.$$

Proof. We choose polyhedral structures $\mathcal{X}'$ of X and $\mathcal{Y}'$ of Y. Then we can write $|Y|$ as polyhedral set

$$|Y| = \bigcup_{\sigma' \in \mathcal{Y}'} \sigma' \cup \bigcup_{\tau' \in \mathcal{X}'} f(\tau').$$

Applying the construction of lemma 1.1.5 to this union, we get a subdivision of V'

$$\mathcal{H} = \mathcal{H}_{\alpha_1} \cap \ldots \cap \mathcal{H}_{\alpha_n},$$

where the α_i are all affine linear forms occurring in the description of all polyhedra in the above union (as equations $\alpha_i(x) \geq 0$). This leads to the polyhedral structure

$$\mathcal{Y} := \mathcal{Y}' \cap \mathcal{H} = \{\sigma \in \mathcal{H} | \sigma \subseteq |Y|\}.$$

Moreover, according to the previous remark 1.3.2, we can form the polyhedral structure

$$\mathcal{X} := \mathcal{X}' \cap \mathcal{H}_{\alpha_1 \circ f} \cap \ldots \cap \mathcal{H}_{\alpha_n \circ f}.$$

We show that these polyhedral structures satisfy the statement of the lemma. A cell of $\mathcal{X}$ has the form

$$\tau = \{x \in \tau' | \alpha_i(f(x)) \geq 0, \alpha_j(f(x)) = 0, \alpha_k(f(x)) \leq 0$$
$$\text{for all } i \in I, j \in J, k \in K\},$$

where $\tau' \in \mathcal{X}'$ and I, J, K form a partition of $\{1, \ldots, n\}$. It follows that the image is

$$f(\tau) = \{y \in f(\tau') | \alpha_i(y) \geq 0, \alpha_j(y) = 0, \alpha_k(y) \leq 0$$
$$\text{for all } i \in I, j \in J, k \in K\}.$$

Now $f(\tau')$ itself can be written in the form $f(\tau') = \{y \in V' | \alpha_s(y) \geq 0 \text{ for all } s \in S\}$, where $S \subseteq \{1, \ldots, n\}$ is a suitable index set. It follows that

$$f(\tau) = \{y \in V' | \alpha_i(y) \geq 0, \alpha_j(y) = 0, \alpha_k(y) \leq 0$$
$$\text{for all } i \in I', j \in J', k \in K'\}$$

for an adapted partition I', J', K' of $\{1, \ldots, n\}$. Now this is a cell of $\mathcal{H}$ by definition, and contained in $|Y|$, of course. Therefore $f(\tau) \in \mathcal{Y}$, which proves the claim. $\qquad\square$

Example 1.3.5
In example 1.3.3, the polyhedral structures $\mathcal{L}_1^2$ and $\mathcal{L}_1^1$ are appropriate for both morphisms f_1 and f_2.

1.3.2 Push forwards of cycles

Definition and Lemma 1.3.6 (Push forwards of tropical cycles)
Let $f : X \to Y$ be a morphism and choose polyhedral structures $\mathcal{X}$ and

$\mathcal{Y}$ as in the previous lemma. We equip the polyhedral subcomplex of $\mathcal{Y}$

$$f_*\mathcal{X} := \{f(\tau)|\tau \in \mathcal{X} \text{ contained in a facet of } \mathcal{X} \text{ on which } f \text{ is injective}\}$$

with the weight function

$$\omega_{f_*\mathcal{X}}(\sigma') = \sum_{\substack{\sigma \in \mathcal{X} \\ f(\sigma)=\sigma'}} \omega_{\mathcal{X}}(\sigma) \cdot [\Lambda'_{\sigma'} : f_\sigma(\Lambda_\sigma)]. \tag{1.7}$$

We claim that this is a balanced polyhedral complex and call the associated tropical cycle the *push forward* f_*X of X. Note that this definition is indeed well-defined: If we start with refinements of $\mathcal{X}$ and $\mathcal{Y}$ (and apply the previous lemma), we obtain a refinement of $f_*\mathcal{X}$ with corresponding weights. Note that by definition $|f_*X| \subseteq f(|X|)$ holds. In particular, if $f(|X|)$ is of strictly lower dimension than X, then $f_*X = \emptyset$.

For a subcycle $Z \in Z_*(X)$, we define the *push forward of Z by $f_*Z :=$* $(f|_Z)_*Z$. This provides a group homomorphism

$$f_* : Z_*(X) \to Z_*(Y).$$

Proof. We want to prove that $f_*\mathcal{X}$ is balanced and that $f_* : Z_*(X) \to Z_*(Y)$ is a group homomorphism. The latter one is straightforward: Let $Z, Z' \in Z_d(X)$ be two subcycles of dimension d. We can choose polyhedral structures $\mathcal{Z}$ of $|Z| \cup |Z'|$ and $\mathcal{Y}$ of Y according to lemma 1.3.4. Then Z, Z' and $Z + Z'$ are given by weight functions ω_Z, $\omega_{Z'}$ and $\omega_{Z+Z'}$ on $\mathcal{Z}$ which satisfy $\omega_Z + \omega_{Z'} = \omega_{Z+Z'}$. As the weight formula 1.7 is linear with respect to the weights of the source cycle, the claim follows.

So let us prove that $f_*\mathcal{X}$ is balanced. A similar proof can be found in [GKM07, proposition 2.25]. First we want to prove an appropriate locality statement, that will also be helpful in what follows. Again, let $\mathcal{X}$ and $\mathcal{Y}$ be polyhedral structures as in lemma 1.3.4 and take a cell $\tau \in \mathcal{X}^{(e)}$ such that $f(\tau) = \tau' \in \mathcal{Y}^{(e)}$ is of the same dimension. This implies that the integer linear part of f at τ maps V_τ to $V'_{\tau'}$ and therefore induces an integer linear map $f^\tau : V_{U(\tau)}/V_\tau \to V'/V'_{\tau'}$. Therefore we can formulate the following statement:

Lemma 1.3.7
Let τ' be an e-dimensional cell of $f_*\mathcal{X}$. Then the local formula

$$\mathrm{Star}_{f_*\mathcal{X}}(\tau') = \sum_{\substack{\tau \in \mathcal{X}^{(e)} \\ f(\tau) = \tau'}} [\Lambda'_{\tau'} : f_\tau(\Lambda_\tau)] \cdot f^\tau_*(\mathrm{Star}_\mathcal{X}(\tau))$$

holds.

Proof. It follows from $f^\tau(\bar\sigma) = \overline{f(\sigma)}$ that also the right hand side can be computed on the polyhedral complex $\mathrm{Star}_{f_*\mathcal{X}}(\tau')$, i.e. we only have to compare the weights.

By the weight formula 1.7, the weight of a facet $\bar\sigma' \in \mathrm{Star}_{f_*\mathcal{X}}(\tau')$ is a sum over all facets $\sigma \in \mathcal{X}$ with $f(\sigma) = \sigma'$ on the left hand side and over all $\tau < \sigma$ with $f(\tau) = \tau'$ and $f^\tau(\bar\sigma) = \bar\sigma'$ on the right hand side. These two index sets are actually equivalent, as a facet $\sigma \in \mathcal{X}$ with $f(\sigma) = \sigma' > \tau'$ contains one unique face $\tau < \sigma$ with $f(\tau) = \tau'$ (as $f|_\sigma$ is supposed to be injective).

So it remains to compare the varying factors appearing in the weight formula, or more precisely, to show the equality

$$[\Lambda'_{\sigma'} : f_\sigma(\Lambda_\sigma)] = [\Lambda'_{\tau'} : f_\tau(\Lambda_\tau)] \cdot [\Lambda'_{\bar\sigma'} : f^\tau(\Lambda_{\bar\sigma})],$$

which follows from the third isomorphism theorem and from $\Lambda'_{\bar\sigma'} = \Lambda'_{\sigma'}/\Lambda'_{\tau'}$ resp. $f^\tau(\Lambda_{\bar\sigma}) = f_\sigma(\Lambda_\sigma)/f_\tau(\Lambda_\tau)$. This proves the claim. $\square$

The previous locality statement shows that $\mathrm{Star}_{f_*\mathcal{X}}(\tau')$ is balanced if all $f^\tau_*(\mathrm{Star}_\mathcal{X}(\tau))$ are balanced. Thus, to finish the proof that $f_*\mathcal{X}$ is balanced, we can restrict to the case where $\mathcal{X}$ is a one-dimensional fan and $f : \Lambda \to \Lambda'$ is an integer linear function. Let $u_\varrho := u_{\varrho/\{0\}}$ be the primitive generator of the ray. Now if we apply f to the balancing equation of $\mathcal{X}$, we get

$$\sum_{\varrho \in \mathcal{X}^{(1)}} \omega_{\mathcal{X}}(\varrho) f(u_\varrho) = 0$$

Of course we can omit the rays with $f(u_\varrho) = 0$. For the others, the formula $f(u_\varrho) = [\Lambda'_{f(\varrho)} : f(\Lambda_\varrho)] u_{f(\varrho)}$ holds. Plugging this in and rearranging the sum accordingly, we get

$$\sum_{\varrho' \in f_* \mathcal{X}^{(1)}} \sum_{\varrho : f(\varrho) = \varrho'} \omega_{\mathcal{X}}(\varrho)[\Lambda'_{\varrho'} : f(\Lambda_\varrho)] u_{\varrho'} = 0,$$

which is precisely the balancing condition for $f_* \mathcal{X}$ with weight formula 1.7. $\qquad\square$

Example 1.3.8 • Let us return to example 1.3.3. f_2 contracts τ_2, but maps the cells τ_0 and τ_1 with trivial index onto $\mathbb{R}_{\geq 0}$ and $\mathbb{R}_{\leq 0}$ respectively. Thus we compute $(f_2)_* \mathcal{L}_1^2 = \mathcal{L}_1^1$ and $(f_2)_* L_1^2 = \mathbb{R}$. Now consider f_1. On the one hand, we have two cells τ_1 and τ_2 mapping to $\mathbb{R}_{\leq 0}$, both with index 1. On the other hand, there is only τ_0 mapping to $\mathbb{R}_{\geq 0}$, but with index 2 as $f_1((1,1)) = 2$. So again, $(f_1)_* \mathcal{L}_1^2$ is balanced, but this time the push forward equals $(f_2)_* L_1^2 = 2 \cdot \mathbb{R}$.

• Let $f : V \to W$ and $f' : V' \to W'$ be two integer affine maps of vector spaces with lattices. Then

$$(f \times f')_*(X \times X') = f_*(X) \times f'_*(X')$$

holds. Indeed, after choosing appropriate polyhedral structures of X and X', this follows immediately from the definitions.

Remark 1.3.9 (Pushing forward is functorial)
Note that pushing forward, as we expect, is functorial: The identity map

id : $X \to X$ obviously induces the trivial push forward $\mathrm{id}_*(Z) = Z$ for all subcycles Z. Moreover, let $f : X \to Y$ and $g : Y \to Z$ be two morphisms. When we apply lemma 1.3.4, first to f, then to g and to $g \circ f$, we obtain polyhedral structures $\mathcal{X}, \mathcal{Y}, \mathcal{Z}$ such that $f(\tau) \in \mathcal{Y}$ for all $\tau \in \mathcal{X}$ and $g(\tau') \in \mathcal{Z}$ for all $\tau' \in \mathcal{Y}$. Then the weights of facet σ'' in $(g \circ f)_*(\mathcal{X})$ and $g_*(f_*(\mathcal{X}))$ are both computed as a sum over the facets $\sigma \in \mathcal{X}$ with $(g \circ f)(\sigma) = g(f(\sigma)) = \sigma''$ and the occurring factors (with $f(\sigma) = \sigma'$)

$$[\Lambda_{\sigma''} : g_{\sigma'}(f_\sigma(\Lambda_\sigma))] = [\Lambda_{\sigma''} : g_{\sigma'}(\Lambda_{\sigma'})] \cdot [g_{\sigma'}(\Lambda_{\sigma'}) : g_{\sigma'}(f_\sigma(\Lambda_\sigma))]$$
$$= [\Lambda_{\sigma''} : g_{\sigma'}(\Lambda_{\sigma'})] \cdot [\Lambda_{\sigma'} : f_\sigma(\Lambda_\sigma)]$$

are also equal. Hence the equality $(g \circ f)_* X = g_*(f_* X)$ follows.

1.3.3 The projection formula

Definition 1.3.10 (Pull backs of Cartier divisors)
Let $f : X \to Y$ be a tropical morphism and let $\varphi \in \mathrm{Rat}(Y)$ be a rational function on Y. Then we define the *pull back $f^*\varphi$ of φ* to be the rational function $\varphi \circ f$ on X. To check that $\varphi \circ f$ is a rational function, we can apply the construction of lemma 1.3.4 to a polyhedral structure $\mathcal{Y}'$ on whose cells φ is affine.

Moreover, if $\psi \in \mathrm{Div}(Y)$ is a Cartier divisor on Y with representative $\{(U_i, \psi_i)\}$, then we define the *pull back $f^*\psi$ of ψ* to be the Cartier divisor on X given by

$$\{(f^{-1}(U_i), \psi_i \circ f)\}.$$

Again, we can use lemma 1.3.4 with respect to a sufficiently fine polyhedral structure $\mathcal{Y}'$ of Y to show that this indeed defines a Cartier divisor. The definition is independent of the chosen representative as f is locally affine and therefore $(\psi_i \circ f)|_{U(\tau)}$ is affine if ψ_i is affine.

Note that the pull back of a convex Cartier divisor/rational function is convex again as the composition of a convex function and a linear map is again convex.

We can now state and prove the important projection formula.

Theorem 1.3.11 (Projection formula)
Let $f : X \to Y$ be a tropical morphism and let $\varphi \in \text{Div}(Y)$ be a Cartier divisor on Y. Then the following formula holds:

$$\varphi \cdot f_* X = f_*(f^* \varphi \cdot X) \in Z_{\dim X - 1}(Y)$$

Proof. Let $\{(U_i, \varphi_i)\}$ be a representative of φ. We choose polyhedral structures $\mathcal{X}$ and $\mathcal{Y}$ such that $\{f(\tau) | \tau \in \mathcal{X}\} \subseteq \mathcal{Y}$, the U_i are polyhedral with respect to $\mathcal{Y}$ and the φ_i are affine on the cells of $\mathcal{Y}$ (where defined). This implies that the respective statements also hold for $\mathcal{X}$ and $\{(f^{-1}(U_i), \varphi_i \circ f)\}$.

Now the claim is an easy consequence of the locality of the involved operation. We have to compare the weights of a cell $\tau' \in \mathcal{Y}^{(\dim X - 1)}$ on both sides. To do this, we apply the locality formulas for the intersection products and the push forwards on both sides (note that taking germs of φ commutes with pulling back). Therefore we can restrict to the case where $\mathcal{X}$ and $\mathcal{Y} := f_* \mathcal{X}$ are one-dimensional fans, $f : \Lambda \to \Lambda'$ is integer linear, and φ is a fan function on $f_* \mathcal{X}$. In this case, a direct computation shows

$$\begin{aligned}
\omega_{f_*(f^* \varphi \cdot \mathcal{X})}(\{0\}) &= \omega_{f^* \varphi \cdot \mathcal{X}}(\{0\}) \\
&= \sum_{\varrho \in \mathcal{X}} \omega_{\mathcal{X}}(\varrho) \varphi(f(u_\varrho)) \\
&= \sum_{\varrho' \in f_* \mathcal{X}} \sum_{\varrho : f(\varrho) = \varrho'} \omega_{\mathcal{X}}(\varrho)[\Lambda'_{\varrho'} : f(\Lambda_\varrho)] \varphi(u_{\varrho'}) \\
&= \sum_{\varrho' \in f_* \mathcal{X}} \omega_{f_* \mathcal{X}}(\varrho) \varphi(u_{\varrho'}) \\
&= \omega_{\varphi \cdot f_* \mathcal{X}}(\{0\}),
\end{aligned}$$

which proves the claim. □

Example 1.3.12

As an example of the usefulness of the projection formula, let us mention
how it simplifies the proof of lemma 1.2.9. Let $h_1, \ldots, h_r$ be integer linear
functions on a vector space V of dimension r, and consider the rational
functions $\max\{h_i, 0\}$ on V. Then the linear map $H : V \rightarrow \mathbb{R}^r$ given
by $H(x) = (h_1(x), \ldots, h_l(x))$ is a tropical morphism and $\max\{h_i, 0\} =
H^*(\max\{x_i, 0\})$, where x_i denotes the i-th standard coordinate function
on $\mathbb{R}^r$. Hence with help of the projection formula and the fact that the
push forward preserves the degree of a zero-dimensional cycle (see section
1.4, in particular remark 1.4.8), we can conclude

$$\deg(\max\{h_1, 0\} \cdots \max\{h_r, 0\} \cdot V)$$
$$= \deg(\max\{x_1, 0\} \cdots \max\{x_r, 0\} \cdot H_*(V)).$$

But then, we furthermore know $H_*(V) = [\mathbb{Z}^r : H(\Lambda)] \cdot \mathbb{R}^r$ by definition
(assuming H has full rank) and can easily compute

$$\deg(\max\{x_1, 0\} \cdots \max\{x_r, 0\} \cdot \mathbb{R}^r) = 1.$$

Thus, the claim of lemma 1.2.9 is verified in the special case $r = l$ and
can easily be extended to the general case by locality.

1.4 Rational equivalence

Up to now, there was no need of introducing a notion of rational equivalence, as the definition of the intersection product of Cartier divisors
and tropical cycles works without it. However, for later applications in
enumerative geometry it is good to have a tool that guarantees numerical
invariance when varying e.g. the point configurations which the counted

curves have to meet. This is why we define a notion of rational equivalence here that satisfies numerical invariance with respect to intersection products and covers for example translations of cycles. Moreover, we give a description of the tropical Chow group of a vector space V. The material is mostly combined from section 8 of [AR07] and from [AR08].

1.4.1 The Picard group

The first thing we need, in order to establish a concept of rational equivalence, is an appropriate class of functions that generates the equivalence. In classical algebraic geometry, this is basically the class of all rational functions. This seems to be too much in tropical geometry, as this would make many Picard groups and Chow groups, say of a tropical line $L = \max\{0, x, y\} \cdot \mathbb{R}^2$, trivial, which we do not want forcertain reasons. The problem arises from the fact that tropical cycles, as we defined them, are usually non-compact (if not zero-dimensional). This leads, as in the classical case, to rational function on X, which "have hidden zeros and poles" in the "boundary" of X. Consider for example the function $\varphi = \max\{0, x, y\}|_L$ on L again. Let us define the *degree* $\deg(X)$ *of a zero-dimensional cycle* X to be the sum of all weights, i.e. $\deg(X) = \sum_{p \in X} \omega_X(P)$. Then the Weil divisor of φ is equal to the vertex of L with weight 1 and therefore has degree one. On the other hand, on the ray $\mathbb{R}_{\geq 0}(1, 1)$ the slope of the function is 1, so we would expect that the function has a pole at the infinite point of this ray. As this point does not belong to L, this pole is not within our scope. But if rational equivalence did not preserve the degree, it would be quite useless in applications (e.g. in enumerative geometry). Therefore we would like to generate rational equivalence by such functions "whose divisor in any compactification has no components in the boundary". This can be achieved by restricting to bounded functions.

Remark 1.4.1

There is another possibility how to deal with the non-compactness of (sub-cycles of) V. Namely, one can try to compactify the cycles in question and then extend the intersection theory to such compact objects. However, this approach is not treated in this thesis. Instead, this is part of active research.

Definition 1.4.2 (Rational equivalence generated by bounded functions)

Let Z be an abstract tropical cycle and $R(Z) := \{[(|Z|, \phi)] | \phi \text{ bounded}\}$ be the group of all Cartier divisors globally given by a bounded rational function. We define the *Picard group of Z* to be the quotient $\mathrm{Pic}(Z) := \mathrm{Div}(Z)/R(Z)$. We call two Cartier divisors (rational functions) *rationally equivalent*, if their classes in $\mathrm{Pic}(Z)$ are the same. In particular, two rationally equivalent rational functions differ in a sum of a bounded and a locally affine function.

Remark 1.4.3

Let $f : X \to Y$ be a tropical morphism and ϕ on Y be a bounded rational function. Then also $f^*\phi$ is bounded and therefore we get a well-defined pull back homomorphism

$$f^* : \mathrm{Pic}(Y) \to \mathrm{Pic}(X).$$

Let us now prove that we do not divide out too much for applications in enumerative geometry.

Lemma 1.4.4

Let X be a one-dimensional tropical cycle, ϕ a bounded rational function on X. Then the degree of $\mathrm{div}(\phi)$ is zero, $\deg(\mathrm{div}(\phi)) = 0$.

Proof. Let $\mathcal{X}$ be a polyhedral structure on whose cells ϕ is affine. Then by definition for all $\{p\} \in \mathcal{X}^{(0)}$ we have

$$\omega_{\mathrm{div}(\phi)}(\{p\}) = \sum_{\substack{\sigma \in \mathcal{X}^{(1)} \\ p \in \sigma}} \omega(\sigma)\phi_\sigma(u_{\sigma/\{p\}}),$$

and therefore

$$\deg(\mathrm{div}(\phi)) = \sum_{p,\sigma} \omega(\sigma)\phi_\sigma(u_{\sigma/\{p\}}),$$

where the sum runs through all flags $\mathcal{X}^{(0)} \ni \{p\} < \sigma \in \mathcal{X}^{(1)}$. Now, if $\sigma \in X^{(1)}$ contains two different vertices, say $\partial_1\sigma$ and $\partial_2\sigma$, then the two corresponding primitive generators have opposite directions, i.e. $u_{\sigma/\{\partial_1\sigma\}} = -u_{\sigma/\{\partial_2\sigma\}}$. Hence the two corresponding terms in the above sum cancel out. If, otherwise, σ contains no two vertices, then σ must be an unbounded polyhedron and therefore ϕ can only be bounded if it is constant on σ. Therefore $\phi_\sigma \equiv 0$ and the claim follows. $\qquad\square$

1.4.2 Chow groups

Definition 1.4.5 (Rational equivalence of cycles)
Let Z be a cycle and let X be a subcycle. We call X *rationally equivalent to zero on Z*, denoted by $X \sim 0$, if there exists a morphism $f : Z' \to Z$ from an arbitrary further tropical cycle Z' to Z and a bounded rational function ϕ on Z' such that

$$f_*(\phi \cdot Z') = X.$$

Let Y be another subcycle of Z. Then we call X and Y *rationally equivalent* if $X - Y$ is rationally equivalent to zero, denoted by $X \sim Y$. The group of equivalence classes $A_*(X) := Z_*(X)/ \sim$ is called the *Chow group of X*.

Remark 1.4.6

The push forward construction is necessary as otherwise push forwards were not compatible with rational equivalence (cf. [AR07, remark 8.6]). Moreover, the definition given here still satisfies all desired properties, as we will see in the following.

Lemma 1.4.7

Let X be a cycle in Z rationally equivalent to zero. Then the following holds:

(a) *Let φ be a Cartier divisor on Z. Then $\varphi \cdot X$ is also rationally equivalent to zero.*

(b) *Let $g : Z \to \widetilde{Z}$ be a morphism to an arbitrary further tropical cycle $\widetilde{Z}$. Then $g_*(X)$ is also rationally equivalent to zero.*

(c) *Assume that X is zero-dimensional. Then $\deg(X) = 0$.*

Proof. Let $f : Z' \to Z$ be a morphism and ϕ a bounded function on Z' such that $f_*(\phi \cdot Z') = X$. Then restricting f to $f : f^*\varphi \cdot Z' \to Z$ and the projection formula provide (a), whereas composing f with g provides (b). For (c), Z' must be one-dimensional and we can apply 1.4.4, which shows that the degree of $\phi \cdot Z'$ is zero. But pushing forward preserves the degree. $\qquad\square$

Remark 1.4.8

The previous lemma guarantees the existence of

- a bilinear intersection product

$$\cdot : \operatorname{Pic}(X) \times A_*(X) \to A_{*-1}(X),$$

- for every tropical morphism $f : X \to Y$ a Chow group homomorphism

$$f_* : A_*(X) \to A_*(Y),$$

- and a well-defined degree homomorphism

$$\deg : A_0(X) \to \mathbb{Z}.$$

An easy example of rationally equivalent cycles are translations.

Lemma 1.4.9 (Translations are rationally equivalent)
Let X be a cycle in $\mathbb{R}^r$ and let $X(v)$ denote the translation of X by an arbitrary vector $v \in \mathbb{R}^r$. Then the equation

$$X(v) \sim X$$

holds.

Proof. Consider the cycle $X \times \mathbb{R}$ in $\mathbb{R}^r \times \mathbb{R}$ with morphism

$$f : \mathbb{R}^r \times \mathbb{R} \to \mathbb{R}^r,$$
$$(x, t) \mapsto x + t \cdot e_i,$$

where e_i is the i-th unit vector in $\mathbb{R}^r$ (note that we cannot use v here, as v might not have integer slope, hence f might not be integer affine). For $\mu \in \mathbb{R}_{\geq 0}$ let ϕ_μ be the bounded function

$$\phi_\mu(x, t) = \begin{cases} 0 & t \leq 0 \\ t & 0 \leq t \leq \mu \\ \mu & t \geq \mu. \end{cases}$$

Then $f_*(\phi_\mu \cdot X \times \mathbb{R}) = X - X(\mu \cdot e_i)$, which shows that $X \sim X(\mu \cdot e_i)$ holds. Hence the claim follows by iteration. $\square$

1.4.3 The recession cycle

The following definitions and statements can be found in [Zi94] (cf. definition 1.11 and proposition 1.12). Let τ be a (non-empty) polyhedron in V. We define the *recession cone of τ* to be

$$\mathrm{rc}(\tau) := \{v \in V \,|\, x + \mathbb{R}_{\geq 0} v \subseteq \tau \forall x \in \tau\} = \{v \in V \,|\, \exists x \in \sigma \text{ s.t. } x + \mathbb{R}_{\geq 0} v \subseteq \tau\}.$$

The two sets coincide as τ is closed and convex. If τ' is another polyhedron with non-empty intersection $\tau \cap \tau' \neq \emptyset$, then $\mathrm{rc}(\tau \cap \tau') = \mathrm{rc}(\tau) \cap \mathrm{rc}(\tau')$ holds. Let us denote the set of vertices of τ by $\tau^{(0)}$. If τ has at least one vertex, the formula

$$\tau = \mathrm{conv}(\tau^{(0)}) + \mathrm{rc}(\tau) \tag{1.8}$$

holds, where conv denotes the convex hull. One can easily conclude that for every face $\tau' < \tau$, the recession cone $\mathrm{rc}(\tau')$ is a face of $\mathrm{rc}(\tau)$. Moreover, every face of $\mathrm{rc}(\tau)$ is of this form.

Lemma 1.4.10

Let X be a tropical cycle in V. Then there exists a polyhedral structure $\mathcal{X}$ such that

$$\{\mathrm{rc}(\tau) \,|\, \tau \in \mathcal{X}\}$$

is a fan.

Proof. We start with an arbitrary polyhedral structure $\mathcal{X}'$ such that every cell contains vertices (otherwise we intersect such a cell with a translated complete fan). Again, we apply the construction of lemma 1.1.5 to the polyhedral set $\bigcup_{\tau' \in \mathcal{X}} \mathrm{rc}(\tau')$ and obtain a complete fan $\mathcal{H}$ such that all $\mathrm{rc}(\tau')$ are unions of cells of $\mathcal{H}$. Then we form $\mathcal{X} := \mathcal{X}' \cap \mathcal{H}$ and consider a (non-empty) cell $\tau = \tau' \cap \sigma, \tau \in \mathcal{X}, \sigma \in \mathcal{H}$. Its recession cone is $\mathrm{rc}(\tau) = \mathrm{rc}(\tau') \cap \mathrm{rc}(\sigma)$, which is a cone in $\mathcal{H}$ by construction. This proves the claim. $\square$

Definition 1.4.11 (The degree of tropical cycles)

Let X be a d-dimensional tropical cycle in V, let $\mathcal{X}$ be a polyhedral structure such that

$$\{\mathrm{rc}(\tau)|\tau \in \mathcal{X}\}$$

is a fan. The pure-dimensional fan which is obtained by removing all cones that are not contained in a d-dimensional facet is denoted by $\delta(\mathcal{X})$. To every facet $\sigma \in \delta(\mathcal{X})^{(d)}$ we assign the weight

$$\omega_{\delta(\mathcal{X})}(\sigma) := \sum_{\substack{\sigma' \in \mathcal{X} \\ \mathrm{rc}(\sigma')=\sigma}} \omega_X(\sigma').$$

The associated tropical fan (see proof below) is called the *degree of X* and is denoted by $\delta(X)$. It is independent of the chosen polyhedral structure, as a refinement of $\mathcal{X}$ leads to a compatible refinement of $\delta(\mathcal{X})$.

Before we prove that $\delta(\mathcal{X})$ is balanced, let us state an important property of $\delta(X)$.

Theorem 1.4.12

Let X be a cycle in V. Then X is rationally equivalent (in V) to its degree $\delta(X)$, i.e.

$$X \sim \delta(X).$$

As the proof of this statement is to a large extent the work of Lars Allermann, my coauthor of [AR07] and [AR08], we skip it here (see [AR08, theorem 7]). Instead, we prove explicitly that $\delta(\mathcal{X})$ is balanced (which in [AR08] is proven only implicitly as part of the proof of theorem 7).

Proof of the balancing condition of $\delta(\mathcal{X})$. Let us first assume $\dim(\mathcal{X}) = 1$. Adding up the balancing condition for all vertices $\nu \in \mathcal{X}^{(0)}$, we get

$$\sum_{V,\sigma} \omega_X(\sigma) u_{\sigma/\nu} = 0,$$

where the sum is subject to all flags $\mathcal{X}^{(0)} \ni \nu < \sigma \in \mathcal{X}^{(1)}$. Every edge σ with $\mathrm{rc}(\sigma) = \{0\}$ is the convex hull of two vertices ν, ν' with $u_{\sigma/\nu} = -u_{\sigma/\nu'}$ and hence its contribution to the sum cancels out. What remains is precisely the balancing condition for $\delta(\mathcal{X})$, as obviously $u_{\sigma/\nu} = u_{\mathrm{rc}(\sigma)/\{0\}}$ holds if $\mathrm{rc}(\sigma)$ is a ray.

For the general case, let τ' be a ridge of $\delta(\mathcal{X})$ and consider the polyhedral open set

$$U := \bigcup_{\substack{\sigma \in \mathcal{X} \\ \tau' \subseteq \mathrm{rc}(\sigma)}} \mathrm{RelInt}(\sigma).$$

Then $V_{\tau'}$ is contained in the common space of linearily of U and it is easy to check that the image of U under the quotient map $q : V \to V/V_{\tau'}$ induces a one-dimensional tropical cycle whose degree equals $\mathrm{Star}_{\delta(\mathcal{X})}(\tau')$. Indeed, let $\mathcal{X}_{\tau'}$ denote the set of all cells $\sigma \in \mathcal{X}$ with $\tau' \subseteq \mathrm{rc}(\sigma)$. The inclusion $q(\sigma) \subseteq q(U)$ is true for all $\sigma \in \mathcal{X}_{\tau'}$, as for every $p \in \sigma$ the subset $p + \tau'$ of σ intersects U for dimensional reasons. Therefore the properties of recession cones mentioned at the beginning of this subsection ensure that $\mathcal{Y} := \{q(\sigma) | \sigma \in \mathcal{X}_{\tau'}\}$ forms a polyhedral complex — if we furthermore show that $q(\sigma) \cap q(\sigma') \neq \emptyset$ implies $\sigma \cap \sigma' \neq \emptyset$ for all $\sigma, \sigma' \in \mathcal{X}_{\tau'}$. So assume $p \in \sigma, p' \in \sigma'$ with $q(p) = q(p')$. It follows that the affine space $p + V_{\tau'}$ contains the two "translated" full-dimensional cones $p + \tau'$ and $p' + \tau'$, which must intersect (look at them "from far away"). As $p + \tau' \subseteq \sigma, p' + \tau' \subseteq \sigma'$, this implies that σ and σ' intersect. With weights $\omega_{\mathcal{Y}}(q(\sigma)) := \omega_{\mathcal{X}}(\sigma)$ and the fact that the star around the ridge $\tau \in \mathcal{X}_{\tau'}$ and the star around $q(\tau)$ are equal, it follows that $\mathcal{Y}$ is a balanced polyhedral complex. As moreover "mapping to the quotient" and "taking recession cones" commute, the claim follows.

Then the one-dimensional case shows that $\mathrm{Star}_{\delta(\mathcal{X})}(\tau')$ is balanced and thus $\delta(\mathcal{X})$ is balanced. $\qquad\square$

1.4.4 The Chow groups of a vector space

In order to compute $A_*(V)$, one important ingredient is theorem 1.4.12 which states that any tropical cycle is rationally equivalent to a tropical fan. In other words, let $Z_*^{\mathrm{fan}}(V)$ denote the group of tropical fans in V, then the class map $Z_*^{\mathrm{fan}}(V) \to A_*(V)$ is surjective. We now show that this map is in fact also injective, hence an isomorphism. Before we prove this, we introduce the helpful notion of numerical equivalence.

Definition 1.4.13 (Numerical equivalence)
Let X be a d-dimensional tropical cycle in V. Then we define d_X to be the map

$$d_X : \mathrm{Pic}(V)^d \to \mathbb{Z},$$
$$(\varphi_1, \dots, \varphi_d) \mapsto \deg(\varphi_1 \cdots \varphi_d \cdot X).$$

Let Y be another d-dimensional tropical cycle in V. We call X and Y *numerically equivalent* if $d_X = d_Y$.

Remark 1.4.14
The obvious implication is: If X and Y are rationally equivalent, then they are also numerically equivalent, i.e.

$$X \sim Y \Rightarrow d_X = d_Y.$$

This follows from lemma 1.4.7.

Proposition 1.4.15
Let X and Y be two tropical fans V. Then, if X and Y are numerically equivalent, they actually coincide, i.e.

$$d_X = d_Y \Rightarrow X = Y.$$

Proof. It suffices to show the following: If Z is a tropical fan with $d_Z = 0$, then $Z = \emptyset$. To prove this, we choose a fan structure $\mathcal{Z}$ and a complete unimodular fan Ω such that $\mathcal{Z} \subseteq \Omega$. We have to show that the corresponding weight function $\omega_Z : \Omega^{(d)} \to \mathbb{Z}$ is identically zero (under the assumption $d_Z = 0$). So pick $\sigma \in \Omega^{(d)}$ and assume $\sigma = \tau + \varrho$ with $\tau \in \Omega^{(d-1)}, \varrho \in \Omega^{(1)}$. As Ω is unimodular, we can form the rational function $\varphi := \varphi_\varrho$ which takes value 1 at the primitive generator of ϱ and is identically zero on all other rays of Ω (see example 1.2.2 (e)). Then obviously the weight of τ in $\varphi \cdot Z$ is

$$\omega_{\varphi \cdot Z}(\tau) = \omega_Z(\sigma)\varphi(u_\varrho) = \omega_Z(\sigma).$$

But then, $\varphi \cdot Z$ also fulfills the assumption, namely $d_{\varphi \cdot Z} = 0$. Thus we use induction on d, as $\varphi \cdot Z = \emptyset$ implies $\omega_{\varphi \cdot Z}(\tau) = \omega_Z(\sigma) = 0$ and therefore $Z = \emptyset$. It remains to note that for $d = 0$ the situation is trivial, as then d_Z is constant with value $\deg(Z)$. $\qquad\square$

Let us summarize our results:

Theorem 1.4.16 (The Chow group of vector spaces)
Let X, Y be two tropical cycles in V. Then the following are equivalent:

(a) $X \sim Y$

(b) $d_X = d_Y$

(c) $\delta(X) = \delta(Y)$

In other words: The concepts of rational equivalence, numerical equivalence and "having the same degree" coincide. In particular, the class homomorphism provides an isomorphism

$$Z_*^{fan}(V) \cong A_*(V).$$

Proof. (a) $\Rightarrow$ (b) follows from lemma 1.4.7, as mentioned above. (c) $\Rightarrow$ (a) follows from $X \sim \delta(X) = \delta(Y) \sim Y$. Finally, (b) $\Rightarrow$ (c) follows from $d_X = d_{\delta(X)}$ and $d_{\delta(Y)} = d_Y$ (by theorem 1.4.12) and proposition 1.4.15. $\square$

1.5 Intersection products of cycles

So far we are able to intersect Cartier divisors with cycles. Our aim in this section is now to define the intersection of two subcycles inside some vector space $V = \Lambda \otimes \mathbb{R}$. We prove that this intersection product is associative, bilinear and commutative (published in [AR07]). Moreover, we prove a general Bézout-style theorem (published in [AR08]). Finally, we show that in the case of tropical fans this intersection product is equivalent to the fan displacement rule given in [FS94] in order to compute the cup-product of cohomology classes (published in [R08]).

1.5.1 Intersecting tropical cycles

In the following we fix a vector space $V = \Lambda \otimes \mathbb{R}$ together with a set of integer linear coordinate functions $x_1, \ldots, x_r$ (i.e. a basis of $\Lambda^\vee$). If we consider powers V^2 or V^3, the "same" coordinate functions on the second (resp. third) factor are denoted by y_i (resp. z_i). We denote by ψ_i the rational function on $V \times V$ given by $\max\{x_i, y_i\}$ and the associated subdivision of $V \times V$ into half-spaces is denoted by $\mathcal{H}_i := \mathcal{H}_{x_i - y_i}$. For an intersection product $\psi_1 \cdots \psi_r \cdot Z$ for some cycle Z in $V \times V$ we use the shorthand notation $\Psi \cdot Z$. Accordingly, we define the subdivision $\mathcal{H} := \mathcal{H}_1 \cap \cdots \cap \mathcal{H}_r$ of $V \times V$ whose minimal cell is the diagonal $\Delta := \{(v, v) | v \in V\}$. Lemma 1.2.9 proves that

$$\Psi \cdot (V \times V) = \psi_1 \cdots \psi_r \cdot (V \times V) = \Delta$$

holds. Finally, we denote the projection of $V \times V$ to the first factor by $\pi : V \times V \to V : (x, y) \mapsto x$. Now we are ready to define the intersection product of two cycles in V. The idea is to intersect their cartesian product with the diagonal in $V \times V$ and then to push down.

Definition 1.5.1 (Intersection products of cycles)
Let X and Y be two tropical cycles in V. Then we define the *intersection product of X and Y* by

$$X \cdot Y := \pi_*(\Psi \cdot (X \times Y)).$$

Remark 1.5.2
Let us collect the simple consequences of this definition.

- The intersection map

$$\cdot : Z_{r-k}(V) \times Z_{r-l}(V) \to Z_{r-k-l}(V).$$

  is bilinear. This follows readily from the linearity of the push forward, of intersection products with Cartier divisors and of "taking cartesian products".

- As $|\Delta| = |\psi_1| \cap \ldots \cap |\psi_r|$ holds, also $|\Psi \cdot (X \times Y)| \subseteq |\Delta|$ is true and we can as well use the second projection. Therefore the definition is symmetric in X and Y and the commutativity

$$X \cdot Y = Y \cdot X$$

  follows.

- The product satisfies $|X \cdot Y| \subseteq |X| \cap |Y|$ (as $|\Psi \cdot (X \times Y)| \subseteq |\Delta| \cap |X \times Y|$).

- Let $\mathcal{X}$ and $\mathcal{Y}$ be polyhedral structures of X and Y and assume

$\mathrm{codim}_V(X) = k, \mathrm{codim}_V(Y) = l$. Then the codimension of $X \cdot Y$ and the expected codimension of $\mathcal{X} \cap \mathcal{Y}$ is $k + l$. But in general, the polyhedral complex $\mathcal{X} \cap \mathcal{Y}$ is of too high dimension (cf. the right hand picture in definition 1.5.14). However, if we remove cells of too high dimension, i.e. if we form

$$(\mathcal{X} \cap \mathcal{Y})^{(\leq r-k-l)} := \{\tau \in \mathcal{X} \cap \mathcal{Y} | \mathrm{codim}_V(\tau) \geq k + l\},$$

then this polyhedral complex with induced weights (possibly zero) gets a polyhedral structure of $X \cdot Y$. In the following we will meet some tools how to compute the weights of this complex directly, without applying the construction above.

- If X and Y are tropical fans, then $X \cdot Y$ is also a tropical fan. In this case, the ψ_i are fan functions on $X \times Y$, hence $\Psi \cdot (X \times Y)$ is a fan. As π is a linear map, the statement follows.

After this rather simple start, some other seemingly self-evident properties of the intersection product need slightly more work.

Proposition 1.5.3

For all cycles $X \in Z_d(V)$, the equation $V \cdot X = X$ holds.

Proof. Let $\mathcal{X}$ be a polyhedral structure of X. Then we intersect the polyhedral structure $\{V\} \times \mathcal{X}$ of $V \times X$ with $\mathcal{H} = \mathcal{H}_1 \cap \ldots \mathcal{H}_r$ (see above) to get a polyhedral structure $\mathcal{Y}$ of $V \times X$ such that the ψ_i are affine on the cells. We want to compute the weight of a facet σ' of $\Psi \cdot \mathcal{Y}$. Let us first assume $\sigma' \not\subseteq \Delta$ (and therefore $U(\sigma') \cap \Delta = \emptyset$). Then it follows from $\Delta = |\psi_1| \cap \ldots \cap |\psi_r|$ that at least one of the functions ψ_i is affine on $U(\sigma')$. Therefore by locality the weight of σ' must be zero. Now assume $\sigma' \subseteq \Delta$. Then by construction $\sigma' = \{(v, v) | v \in \sigma\}$ for a suitable cell of $\mathcal{X}$, and moreover, $U(\sigma') = V \times \mathrm{RelInt}(\sigma)$. Then passing to $\mathrm{Star}_\mathcal{Y}(\sigma')$, whose support is the vector space $V \times V_\sigma/V_{\sigma'}$, we can apply lemma 1.2.9 to the germs $\psi_i^{\sigma'}$ on $\mathrm{Star}_{\mathcal{Y}}(\sigma')$ (note that, for example, $\psi_i^{\sigma'} = \max\{x_i - y_i, 0\}$ and that $x_1 - y_1, \ldots, x_r - y_r$ form a dual lattice basis of $V \times V_\sigma / V_{\sigma'}$, cf. remark 1.2.10). This means the weight of σ' is

$$\omega_{\Psi \cdot \mathcal{Y}}(\sigma') = \omega_{\{V\} \times \mathcal{X}}(V \times \sigma) = \omega_{\mathcal{X}}(\sigma).$$

Thus $\Psi \cdot \mathcal{Y}$ is given by the polyhedral complex

$$\Big\{ \{(v,v)|v \in \sigma\} \text{ for all } \sigma \in \mathcal{X} \Big\}$$

with the above weight formula, and it follows that $\pi_*(\Psi \cdot \mathcal{Y}) = \mathcal{X}$ holds, which proves the claim. □

Another important property of our intersection product is compatibility with the intersection product of Cartier divisors. The following lemma is necessary to prove this property.

Lemma 1.5.4

Let $X \in Z_d(V)$ and $Y \in Z_e(V')$ be tropical cycles, $\varphi \in \mathrm{Div}(X)$ a Cartier divisor on X and $\pi' : X \times Y \to X$ the projection onto the first factor. Then the equation

$$(\varphi \cdot X) \times Y = \pi'^* \varphi \cdot (X \times Y)$$

holds.

Proof. We choose polyhedral structures $\mathcal{X}$ and $\mathcal{Y}$ such that φ is affine on the cells of $\mathcal{X}$ (and therefore $\pi'^* \varphi$ is affine on the cells of $\mathcal{X} \times \mathcal{Y}$). This means that we can compute both sides on the polyhedral structure $\mathcal{X} \times \mathcal{Y}$ and have to compare the weights of a ridge of $\mathcal{X} \times \mathcal{Y}$. We have two types of such ridges:

i) $\tau \times \tau'$ with $\tau \in \mathcal{X}^{(d-1)}, \tau' \in \mathcal{Y}^{(e)}$,

ii) $\tau \times \tau'$ with $\tau \in \mathcal{X}^{(d)}, \tau' \in \mathcal{Y}^{(e-1)}$.

For the second type we find that the neighbourhood $U(\tau \times \tau')$ projects down to $\mathrm{RelInt}(\tau)$ via π' and therefore $\pi'^* \varphi$ is locally affine. Thus $\tau \times \tau'$ gets weight zero. For the first type, the stars $\mathrm{Star}_{\mathcal{X}}(\tau)$ and $\mathrm{Star}_{\mathcal{X} \times \mathcal{Y}}(\tau \times \tau')$ are isomorphic, as well as the germs φ^{τ} and $(\pi'^* \varphi)^{\tau \times \tau'}$ on it. Thus the claim follows. $\qquad\square$

Corollary 1.5.5

Let $X \in Z_d(V)$ and $Y \in Z_e(V)$ be tropical cycles and $\varphi \in \mathrm{Div}(X)$ a Cartier divisor on X. Then the equation

$$(\varphi \cdot X) \cdot Y = \varphi \cdot (X \cdot Y)$$

holds.

Proof. Let $\pi' : X \times Y \to X$ be the projection onto the first factor. The proof is given by

$$
\begin{aligned}
(\varphi \cdot X) \cdot Y &= \pi'_*(\Psi \cdot [(\varphi \cdot X) \times Y]) \\
\text{(previous lemma 1.5.4)} \quad &= \pi'_*(\pi'^* \varphi \cdot \Psi \cdot [X \times Y]) \\
\text{(projection formula 1.3.11)} \quad &= \varphi \cdot \pi'_*(\Psi \cdot [X \times Y]) \\
&= \varphi \cdot (X \cdot Y).
\end{aligned}
$$

$\qquad\square$

In particular, the previous statement implies that for a cycle given by Cartier divisors, both possibilities to intersect it with other cycles coincide:

Corollary 1.5.6

Let X be a tropical cycle given as an intersection product $X = \varphi_1 \cdots \varphi_k \cdot V$ for suitable Cartier divisors $\varphi_1, \ldots, \varphi_k \in \mathrm{Div}(V)$ and let Y be an arbitrary *cycle in V. Then the equation*

$$\varphi_1 \cdots \varphi_k \cdot Y = X \cdot Y$$

holds. In particular, the intersection is independent of the choice of Cartier divisors describing X.

Proof. Applying lemma 1.5.5 and lemma 1.5.3 we obtain

$$X \cdot Y = (\varphi_1 \cdots \varphi_k \cdot V) \cdot Y = \varphi_1 \cdots \varphi_k \cdot (V \cdot Y) = \varphi_1 \cdots \varphi_k \cdot Y.$$

$\square$

Remark 1.5.7

Note that corollary 1.5.6 also implies that our definition of the intersection product $X \cdot Y$ on V would not change if we used different functions $\psi_1, \ldots, \psi_r$ such that $\psi_1 \cdots \psi_r \cdot (V \times V) = \Delta$. Namely, we can apply corollary 1.5.6 with $\Delta = \Psi \cdot (V \times V)$ and $X \times Y$ inside the vector space $V \times V$.

We now prove that the intersection product of cycles can also be computed locally.

Proposition 1.5.8 (Locality)

Let X, Y be two cycles of codimension k, l in V with polyhedral structures $\mathcal{X}$ and $\mathcal{Y}$. Then the equation

$$\mathrm{Star}_{X \cdot Y}(\tau) = \mathrm{Star}_X(\tau) \cdot \mathrm{Star}_Y(\tau)$$

holds for all polyhedra $\tau \in (\mathcal{X} \cap \mathcal{Y})^{(\leq r-k-l)}$.

Proof. By the previous remark, we can assume that the first $d := \mathrm{codim}_V(\tau)$ elements of our chosen coordinate functions $x_1, \ldots, x_r$ generate $V_\tau^\perp$. Furthermore, let us (by abuse of notation) denote the diagonal map by

$\Delta : V \to V \times V, x \mapsto (x, x)$. By locality of the intersection product with Cartier divisors, we have to compute

$$\text{Star}_{\max\{x_1, y_1\} \cdots \max\{x_r, y_r\} \cdot (X \times Y)}(\Delta(\tau))$$
$$= \max\{x_1, y_1\} \cdots \max\{x_r, y_r\} \cdot \text{Star}_{X \times Y}(\Delta(\tau))$$

and

$$\max\{x_1, y_1\} \cdots \max\{x_d, y_d\} \cdot (\text{Star}_X(\tau) \times \text{Star}_Y(\tau))$$

respectively. Thus the statement follows from the fact that

$$\max\{x_{d+1}, y_{d+1}\} \cdots \max\{x_r, y_r\} \cdot (V \times V/\Delta(V_\tau)) \to V/V_\tau \times V/V_\tau,$$
$$(x, y) \mapsto (x, y)$$

(see lemma 1.2.9 for computation of the source cycle) is an isomorphism and can be restricted to an isomorphism between

$$\max\{x_{d+1}, y_{d+1}\} \cdots \max\{x_r, y_r\} \cdot \text{Star}_{X \times Y}(\Delta(\tau))$$

and $\text{Star}_X(\tau) \times \text{Star}_Y(\tau)$. $\qquad\square$

On our wish list of properties of the intersection product $X \cdot Y$ there is one more item: Associativity.

Proposition 1.5.9 (Associativity)
Let X, Y, Z be tropical cycles in V. Then the equation

$$(X \cdot Y) \cdot Z = X \cdot (Y \cdot Z)$$

holds.

Proof. First, let us fix some notations. We use the projections

$$\pi : V^2 \to V : \qquad (x, y) \mapsto x,$$
$$\pi^{12} : V^3 \to V^2 : \qquad (x, y, z) \mapsto (x, y),$$
$$\pi^{13} : V^3 \to V^2 : \qquad (x, y, z) \mapsto (x, z),$$
$$\pi^1 = \pi \circ \pi^{12} = \pi \circ \pi^{13} : V^3 \to V : \qquad (x, y, z) \mapsto x,$$

and the rational functions

$$\psi_i^{12} = (\pi^{12})^* \psi_i = \max\{x_i, y_i\} : V^3 \to \mathbb{R},$$
$$\psi_i^{13} = (\pi^{13})^* \psi_i = \max\{x_i, z_i\} : V^3 \to \mathbb{R}.$$

We abbreviate the following "products" of rational functions:

$$\Psi^{12} = \psi_1^{12} \cdots \psi_r^{12},$$
$$\Psi^{13} = \psi_1^{13} \cdots \psi_r^{13}.$$

Now we want to prove the following:

$$\pi_*^1 (\Psi^{13} \cdot \Psi^{12} \cdot [X \times Y \times Z]) = (X \cdot Y) \cdot Z.$$

The claim then follows from the fact that exchanging the coordinates y and z (i.e. the second and third factor) does not change the left hand side. So let us prove this equation: We start with the left hand side, replace π^1 by $\pi \circ \pi^{13}$ and use the projection formula to π^{13} and Ψ^{13}, getting

$$\pi_* (\Psi \cdot \pi_*^{13} (\Psi^{12} \cdot [X \times Y \times Z])).$$

Now we use lemma 1.5.4 with Ψ^{12} and $\pi^{12} : V^2 \times V \to V^2$, which provides

$$\pi_* (\Psi \cdot \pi_*^{13} [(\Psi \cdot (X \times Y)) \times Z]).$$

But as $\pi^{13} = \pi \times \mathrm{id}$ and $\pi_*(\Psi \cdot (X \times Y)) = X \cdot Y$, this equals

$$\pi_*(\Psi \cdot [(X \cdot Y) \times Z]) = (X \cdot Y) \cdot Z,$$

which finishes the proof. $\qquad\square$

Remark 1.5.10

One might ask if it is also possible to define intersection products for subcycles of other cycles than V. For our point of view, a necessary condition is that the diagonal of the ambient cycle Z can be expressed in terms of Cartier divisors on $Z \times Z$ (which is in fact very similar to the classical situation). In [Al09], Lars Allermann proves that this condition is satisfied for "smooth" tropical varieties and that an intersection product exists on these varieties. Here "smooth" means that the variety locally looks like (products of) L_d^r for appropriate values $d < r$ (i.e. all stars are of this form).

1.5.2 Bézout's theorem

We now show that our intersection product of cycles is well-defined modulo rational equivalence.

Lemma 1.5.11

Let X and Y be tropical cycles in V resp. V' and assume X is rationally equivalent to zero (in V). Then the following holds:

(a) The cartesian product $X \times Y$ is rationally equivalent to zero (in $V \times V'$).

(b) Assume $V = V'$. Then the intersection product $X \cdot Y$ is also rationally equivalent to zero (in V).

Proof. Let $f : Z \to V$ be a morphism and ϕ a bounded function on Z such that $f_*(\phi \cdot Z) = X$. Then $f \times \mathrm{id} : Z \times V' \to V \times V'$ and $\phi \circ \pi'$

together with lemma 1.5.4 provide (a), where $\pi' : Z \times V' \to Z$ is the first projection.

Assertion (b) follows from

$$X \sim 0$$
$$\text{(part (a))} \quad \Rightarrow \quad (X \times Y) \sim 0$$
$$\text{(lemma 1.4.7 part (a))} \quad \Rightarrow \quad \Psi \cdot (X \times Y) \sim 0$$
$$\text{(lemma 1.4.7 part (b))} \quad \Rightarrow \quad X \cdot Y = \pi_*(\Psi \cdot (X \times Y)) \sim 0$$

$\square$

Corollary 1.5.12

The intersection product on the Chow groups of V

$$\cdot : A_{r-k}(V) \times A_{r-l}(V) \to A_{r-k-l}(V),$$
$$[X] \cdot [Y] \mapsto [X \cdot Y].$$

is well-defined, bilinear and commutative.

Another corollary of this is the following Bézout theorem.

Theorem 1.5.13 (General Bézout's theorem)
Let X, Y be two tropical cycles in V. Then the degrees satisfy

$$\delta(X \cdot Y) = \delta(X) \cdot \delta(Y).$$

Proof. From theorem 1.4.12 and lemma 1.5.11 we know

$$\delta(X \cdot Y) \sim X \cdot Y \sim \delta(X) \cdot \delta(Y).$$

By lemma 1.4.15, two rationally equivalent tropical fans are equal, thus the claim follows. $\square$

1.5.3 The fan displacement rule

In [FS94], the authors compute the cup-product of the toric cohomology groups in terms of Minkowski weights with the help of the so-called fan displacement rule. In this subsection we show explicitly that, when we interpret cohomology classes of compact toric varieties as tropical fans, then the cup-product coincides with our intersection product of tropical cycles. Another approach to this topic is given in [Ka06, section 9]. In order to prove this, we first deal with the case when two cycles X, Y intersect "generically":

Definition 1.5.14 (Transversal intersections)
Let X, Y be two cycles in V of codimension k resp. l. We say X and Y intersect transversally if $|X| \cap |Y|$ is of pure codimension $k + l$ and if for each facet τ in $\mathcal{X} \cap \mathcal{Y}$ (for some polyhedral structures $\mathcal{X}$ and $\mathcal{Y}$) the corresponding neighbourhoods $U_{\mathcal{X}}(\tau)$ and $U_{\mathcal{Y}}(\tau)$ are open sets in (transversal) affine subspaces of V, i.e. if the supports of both $\mathrm{Star}_X(\tau)$ and $\mathrm{Star}_Y(\tau)$ are (transversal) vector spaces.

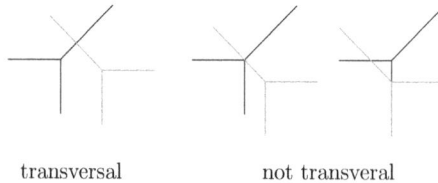

transversal   not transveral

In the case of transversal intersection, by locality of the intersection product the computation of $X \cdot Y$ can be reduced to the intersection of vector spaces.

Lemma 1.5.15 (Intersections of subspaces)
Let U, W be two subspaces of V (with rational slope) such that $U + W = V$. If we consider U, W as cycles with weight 1, their intersection product can

be computed to be

$$U \cdot W = |\Lambda/(\Lambda_U + \Lambda_W)| \cdot (U \cap W).$$

Proof. By definition we have to compute

$$\Psi \cdot (U \times W).$$

In order to use lemma 1.2.9, we use the function $\psi_i = \max\{x_i - y_i, 0\}$ instead of $\max\{x_i, y_i\}$ here. Now we apply lemma 1.2.9. In our case, the function H is just

$$H : \Lambda \times \Lambda \to \Lambda,$$
$$(x, y) \mapsto x - y.$$

Thus $H(\Lambda_U \times \Lambda_W) = \Lambda_U \mp \Lambda_W$ and therefore the restriction to $U \times W$ provides the claim. $\qquad\square$

Now, as a combination of the locality of the intersection product and the previous lemma 1.5.15, we obtain the following result.

Corollary 1.5.16 (Transversal intersections)
Let X, Y be two tropical cycles in V that intersect transversally and let $\mathcal{X}$ and $\mathcal{Y}$ be polyhedral structures. Then $X \cdot Y$ is given by the polyhedral complex $\mathcal{X} \cap \mathcal{Y}$ with the following weight function: Any facet τ in $\mathcal{X} \cap \mathcal{Y}$ is the intersection of two facets σ, σ' in X resp. Y. Then the weight of $\tau = \sigma \cap \sigma'$ is

$$\omega_{X \cdot Y}(\sigma \cap \sigma') = \omega_X(\sigma)\omega_Y(\sigma')|\Lambda/\Lambda_\sigma + \Lambda_{\sigma'}|.$$

Now we turn towards the fan displacement rule. Let Ω be a complete fan in a vector space V. Note that in [FS94], the complete fan is usually called Δ and the exponent in $\Delta^{(k)}$ indicates the codimension, i.e. $\Delta^{(k)}$

means $\Omega^{(r-k)}$ in our notation.

Now let $\gamma \in A^k(\mathbf{X}), \gamma' \in A^l(\mathbf{X})$ be two cohomology classes and let $X(\gamma)$ and $X(\gamma')$ be the associated Ω-directional tropical fans with weight functions ω_γ on $\Omega^{(r-k)}$ and $\omega_{\gamma'}$ on $\Omega^{(r-l)}$. Then the cup-product $\gamma \cup \gamma'$ is a cohomology class of codimension $k + l$ and its associated tropical fan $X(\gamma \cup \gamma')$ is computed in [FS94, 3.1, 3.2]. It is proven that its weight function on $\Omega^{(r-k-l)}$ is given by

$$\omega_{\gamma \cup \gamma'}(\tau) = \sum_{\substack{\sigma \in \Omega^{r-k} \\ \sigma' \in \Omega^{r-l} \\ \tau \subseteq \sigma, \sigma'}} m^\tau_{\sigma,\sigma'} \cdot \omega_\gamma(\sigma) \cdot \omega_{\gamma'}(\sigma').$$

The coefficients $m^\tau_{\sigma,\sigma'}$ appearing in this formula are not unique but depend on the choice of a generic vector $v \in V$. If we fix such a vector v, then

$$m^\tau_{\sigma,\sigma'} = \begin{cases} |\Lambda/\Lambda_\sigma + \Lambda_{\sigma'}| & \text{if } (\sigma + v) \cap \sigma' \neq \emptyset, \\ 0 & \text{otherwise.} \end{cases}$$

The tools introduced in the previous sections make it quite easy now to prove that the cup-product of of cohomology classes is equivalent to our intersection product of tropical cycles in V.

Theorem 1.5.17 (Toric cup-products vs. tropical intersections of cycles)
Let $\gamma, \gamma' \in A^*(\mathbf{X})$ be cohomology classes of codimension k, l. Then the equation

$$X(\gamma) \cdot X(\gamma') = X(\gamma \cup \gamma')$$

holds.

Proof. For each cell $\tau \in \Omega^{(r-k-l)}$ we have to show that the weights of the left and right hand side agree. Note that we can compute both sides locally on $\mathrm{Star}_\Omega(\tau)$. For the left hand side this is possible because of the locality of the intersection product and for the right hand side it corresponds to restricting the cohomology classes to the toric subvariety corresponding to $\text{Star}_\Omega(\tau)$. Since restricting (i.e. pulling back) is compatible with the cup product (cf. [Fu84, section 17.2]), this is allowed as well. Therefore we can assume $k + l = r$ and $\tau = \{0\}$. In this case, by plugging in the definition on the right hand side and choosing a generic vector $v \in V$, it remains to show

$$\deg(X(\gamma) \cdot X(\gamma')) = \sum_{\substack{\sigma \in \Omega^{r-k} \\ \sigma' \in \Omega^{r-l} \\ (\sigma+v) \cap \sigma' \neq \emptyset}} |\Lambda/\Lambda_\sigma + \Lambda_{\sigma'}| \cdot \omega_\gamma(\sigma) \cdot \omega_{\gamma'}(\sigma').$$

Now, for a generic vector $v \in V$ we can assume that all pairs of cells of $\Omega(v)$ and Ω intersect in the expected dimension (in fact, this is basically what the authors of [FS94] mean by a generic vector) and in particular $X(\gamma) + v$ and $X(\gamma')$ intersect transversally. Note that the sum on the right hand side runs through the points in the intersection of $X(\gamma)+v$ and $X(\gamma')$. Therefore, by corollary 1.5.16 it equals $\deg((X(\gamma)+v) \cdot X(\gamma'))$. But as $X(\gamma)+v$ and $X(\gamma)$ are rationally equivalent, the statement follows. $\square$

Remark 1.5.18

Of course, the statement generalizes to stable intersections as suggested in [Mi06] (and discussed for curves in [RGST]). Here, two cycles in $\mathbb{R}^n$ are intersected by applying the fan displacement rule locally (in the sense of proposition 1.5.8). Using the previous results we can conclude that this stable intersection and our intersection product via intersecting with the diagonal coincide. Another approach to that can be found in [Ka09, section 9].

1.6 Tropicalization

The goal of this section is to relate the tropical intersection theory developed so far to the tropicalization of classical varieties via valuation.

The main theorem (not published before) states that tropicalization and complete intersection commute if a certain generic condition is satisfied. This can be regarded as a tropical extension of the Bernshtein bound on the number of solutions of a system of polynomial equations. We get this mainly as a corollary of the properties we discussed before. First, we give a brief account on tropicalization via valuation.

1.6.1 Non-archimedean amoebas

The following can be found in [EKL04] and [SS04]. Let K be an algebraically closed field which is complete with respect to a non-trivial valuation val : $K^* \to \mathbb{R}$, i.e. val satisfies

$$\mathrm{val}(x + y) \geq \min\{\mathrm{val}(x), \mathrm{val}(y)\},$$
$$\mathrm{val}(x \cdot y) = \mathrm{val}(x) + \mathrm{val}(y),$$
$$\mathrm{val} \not\equiv 0.$$

Let $\mathcal{R}$ denote the local ring of elements with non-negative valuation and $\mathfrak{m}$ the maximal ideal of elements with positive valuation. Then the quotient field $\kappa = \mathcal{R}/\mathfrak{m}$ is also algebraically closed (cf. [SS04]). We fix a section $w \mapsto t^w$ from $\mathrm{val}(K^*)$ to K^* such that $t^w \cdot t^v = t^{w+v}$, $\mathrm{val}(t^w) = w$ and $\mathrm{val}(t) = 1$ (w.l.o.g. we assume $1 \in \mathrm{val}(K^*)$). This section t appears naturally as formal variable in the two basic examples, the field of Puiseux series

$$\bigcup_{n \geq 1} \kappa((t^{1/n}))$$

(where κ is an algebraically closed field of characteristic zero), or the field of transfinite Puiseux series $\kappa((t^{\mathbb{Q}}))$ (where κ is an algebraically closed field and the set of exponents of a given formal power series is required to be well-ordered). In both cases, the valuation is given by

$$f = \sum_w a_w t^w \mapsto \min\{w \mid a_w \neq 0\}.$$

However, in the following we assume $\mathrm{val}(K^*) = \mathbb{R}$ to avoid some technicalities.

Now, let V be a purely d-dimensional subvariety of the algebraic torus $T = (K^*)^r$. The *non-archimedean amoeba* $\mathrm{Val}(V)$ *of* V is the image of V under the coordinate-wise valuation map

$$\mathrm{Val} : T \to \mathbb{R}^r,$$
$$(x_1, \ldots, x_r) \mapsto (\mathrm{val}(x_1), \ldots, \mathrm{val}(x_r)).$$

It is proven in [EKL04, theorem 2.2.5] and [BG84] that $\mathrm{Val}(V)$ is a polyhedral set of dimension d which is rational (with respect to $\mathbb{R}^n = \mathbb{Z}^n \otimes \mathbb{R}$).

There are various other descriptions of this set, we mention the following one: Let $f = \sum_I a_I x^I \in K[x_1^\pm, \ldots, x_r^\pm]$ be a non-zero Laurent polynomial and let $\mathrm{NP}(f)$ denote its Newton polytope, i.e. the convex hull of all exponents I with non-zero coefficient a_I. Then for every $w \in \mathbb{R}^r$ we construct a Laurent polynomial $\mathrm{in}_w f \in \kappa[x_1^\pm, \ldots, x_r^\pm]$, called *initial form of* f *with respect to* w. This polynomial is the image of the shifted polynomial $t^C \cdot f(t^w \cdot x)$ under the quotient map $\mathcal{R}[x_1^\pm, \ldots, x_r^\pm] \to \kappa[x_1^\pm, \ldots, x_r^\pm]$, where $t^w = (t^{w_1}, \ldots, t^{w_r}) \in T$ and $C = -\min_{I \in \mathrm{NP}(f)}\{\mathrm{val}(a_I) + \langle I, w \rangle\}$ (i.e. multiplying with t^C makes sure that the result $\mathrm{in}_w f$ is well-defined and never zero, if f is non-zero). Now assume that the subvariety $V \subseteq T$ is given by the ideal J. Then we define the *initial variety* $\mathrm{in}_w V$ to be the subvariety of $T_\kappa = (\kappa^*)^r$ given as the zero set of all polynomials $\mathrm{in}_w f, f \in I$. One can prove that $\mathrm{in}_w V$ is either empty (if one of the $\mathrm{in}_w f$ is a monomial) or d-dimensional. Moreover one can show ([Sp02, theorem 2.1.2]) that $\mathrm{in}_w V$

is non-empty if and only if w is contained in $\mathrm{Val}(V)$, i.e.

$$\mathrm{Val}(V) = \{w \in \mathbb{R}^r \,|\, \mathrm{in}_w(V) \neq \emptyset\}. \tag{1.9}$$

One consequence of this is that if V is defined over κ, then $\mathrm{Val}(V)$ is (the support of) a fan. Let us also mention two (easy) formulas for multiple "in"-taking. The equations

$$\begin{aligned}
\mathrm{in}_v(\mathrm{in}_w f) &= \mathrm{in}_{w+\epsilon v} f \\
\mathrm{in}_v(\mathrm{in}_w V) &= \mathrm{in}_{w+\epsilon v} V
\end{aligned} \tag{1.10}$$

hold for sufficiently small ϵ (cf. [Sp02, proposition 2.2.3]).

1.6.2 Toric degenerations

The next step is to make the polyhedral set $\mathrm{Val}(V)$ into a tropical cycle by defining weights on a suitable polyhedral structure. This can be achieved with the help of toric degenerations. References for this are [Sm96], [NS04], [Sp02] and also [HK08] (it also appears in connection with regular subdivisions of polytopes in Viro's patchworking constructions and in [GKZ94]).

Let Σ be a polyhedral subdivision of $\mathbb{R}^r$, i.e. a polyhedral complex whose support equals $\mathbb{R}^r$. We assume for simplicity that $\{\mathrm{rc}(\tau)|\tau \in \Sigma\} =: \delta(\Sigma)$ is a (complete) fan. To any such Σ we can associate a flat family $\mathbf{X}_{\mathcal{R}}$ over $\mathrm{Spec}(\mathcal{R})$ with the following properties: Recall that $\mathrm{Spec}(\mathcal{R})$ consists of the generic point $\mathrm{Spec}(K)$ and the special point $\mathrm{Spec}(\kappa)$. Then the fiber of the family over the generic point $\mathrm{Spec}(K)$ is the toric variety $\mathbf{X} = \mathbf{X}(\delta(\Sigma))$. The fiber over the special point $\mathrm{Spec}(\kappa)$ is the union of toric varieties $\mathbf{X}_\kappa := \mathbf{X}_\kappa(\Sigma)$ as explained in subsection 1.2.4. So for the example of Σ given there (see page 49), the generic fiber is $\mathbf{X} = \mathbb{P}^1_K \times \mathbb{P}^1_K$, whereas the special fiber is $\mathbf{X}_\kappa$ consists of two copies of $\mathbb{P}^2_\kappa$ with one coordinate axis of each identified.

The important thing about this is that if we fix a purely d-dimensional subvariety $V \subseteq T$, we can choose a suitable subdivision Σ (in particular satisfying $\mathrm{Val}(V) \subseteq |\Sigma^{(d)}|$ and $\mathrm{in}_w(V) = \mathrm{in}_{w'}(V)$ for all $w, w' \in \mathrm{RelInt}(\tau), \tau \in \Sigma$) such that the degeneration of V in the associated family has many nice properties. (Of course, the degeneration V_κ of V is obtained by taking the closure of V in $\mathbf{X}_{\mathcal{R}} \supseteq \mathbf{X} \supseteq T$ and intersecting it with the special fiber $\mathbf{X}_\kappa$.) For example, using $\mathrm{in}_w(V) = \mathrm{in}_{w'}(V)$ for all $w, w' \in \mathrm{RelInt}(\tau), \tau \in \Sigma$ and formula 1.10 one can deduce that for every $w \in \mathrm{RelInt}(\tau)$ the initial variety $\mathrm{in}_w X$ is invariant under the action of the subtorus $T_\tau := \mathrm{Hom}(\mathbb{Z}^r/\Lambda_\tau^\perp, \kappa^*) \subseteq T$. Moreover, the quotient by this action is canonically isomorphic to the orbit intersection $V_\tau = V_\kappa \cap \mathcal{O}_\tau$, i.e.

$$V_\tau \cong \mathrm{in}_w V/T_\tau.$$

In particular, if $\dim(\tau) = d$, then the weight of τ is defined to be the length of the zero-dimensional scheme V_τ, i.e.

$$\omega_{\mathrm{Trop}(V)}(\tau) = \mathrm{length}(V_\tau).$$

It turns out that these weights satisfy the balancing condition and the associated tropical cycle is called the *tropicalization of V*, denoted by $\mathrm{Trop}(V)$. It follows from equation 1.9 that the support of $\mathrm{Trop}(V)$ equals $\mathrm{Val}(V)$. In the special case when V is defined over κ and Σ is a unimodular fan, then $\mathrm{Trop}(V)$ is precisely the tropical fan associated to the cohomology class given by intersecting with the class of the closure of V in $\mathbf{X}$ (or by intersecting with the class of V_κ in $\mathbf{X}_\kappa$).

1.6.3 Generic complete intersections

Let $f = \sum_I a_I x^I \in K[x_1^\pm, \ldots, x_r^\pm]$ be a Laurent polynomial. Then we define the *tropicalization* trop f *of* f by

$$(\text{trop } f)(w) = -\min_{I \in \text{NP}(f)} \{\text{val}(a_I) + \langle I, w \rangle\}$$

(where I is understood to be an integer vector). Note that this defines a tropical polynomial in the sense of example 1.2.2. Taking the minimum here is necessary to be compatible with the valuation theory, but the minus sign makes the tropical polynomial convex — the tropical analogue of "regular". If $f \in \kappa[x_1^\pm, \ldots, x_r^\pm]$, the formula simplifies to

$$(\text{trop } f)(w) = -\min_{I \in \text{NP}(f)} \{\langle I, w \rangle\}. \tag{1.11}$$

Our first result is concerned with the study of the number of solutions of a system of Laurent polynomials.

Theorem 1.6.1 (The tropical Bernshtein bound)
Let $f_1, \ldots, f_r \in \kappa[x_1^\pm, \ldots, x_r^\pm]$ be a system of Laurent polynomials and assume

$$V(\text{in}_w f_1, \ldots, \text{in}_w f_r) = \emptyset$$

for all $w \in \mathbb{R}^r$. Then the number of solutions of this system in T (counted with multiplicities) can be computed tropically by

$$\text{length}(V(f_1, \ldots, f_r)) = \deg(\text{trop } f_1 \cdots \text{trop } f_r \cdot \mathbb{R}^r).$$

Proof. In a three page paper from 1975 [Be75], Bernshtein gives a beautiful interpretation of the number of solutions of a generic system of Laurent polynomials in terms of the associated Newton polytope. Namely under the assumptions of our theorem, the statement is ([Be75, theorem B. a)])

$$\text{length}(V(f_1, \ldots, f_r)) = \text{MV}(\text{NP}(f_1), \ldots, \text{NP}(f_r)),$$

where MV denotes the mixed volume of the Newton polytopes (cf. section 1.2.4). Now let $\mathbf{X}$ be a smooth toric variety whose fan is a refinement of the dual fans of all $\text{NP}(f_i)$. Recall that then the $\text{NP}(f_i)$ define Cartier divisors on $\mathbf{X}$ whose intersection also equals the mixed volume of the polytopes (cf. corollary 1.2.16). Moreover, recall that by formula 1.11 the tropical polynomials $\text{trop } f_i$ agree with the rational functions associated to these Cartier divisors (with a minus sign in contrast to the notation in [Fu93]). Thus the theorem follows from the fact that the tropical product of rational functions on $\mathbb{R}^r$ and the classical product of Cartier divisors on $\mathbf{X}$ are equivalent. $\qquad\square$

Remark 1.6.2
The assumption of the theorem, $V(\text{in}_w f_1, \ldots, \text{in}_w f_r) = \emptyset$ for all $w \in \mathbb{R}^r$, makes sure that in a fine enough toric compactification, the intersection

$$\overline{V(f_1)} \cap \ldots \cap \overline{V(f_r)}$$

does not intersect the boundary, as expected for dimensional reasons. In particular, it implies that $V(f_1, \ldots, f_r)$ is finite.

We can now deal with the more general case of generic complete intersections over K.

Theorem 1.6.3 (Tropicalizations of generic complete intersections)
Let $f_1, \ldots, f_k \in K[x_1^{\pm}, \ldots, x_r^{\pm}]$ be Laurent polynomials and assume that $V := V(f_1, \ldots, f_k)$ is a complete intersection, i.e. $d := \dim(V) = r - k$. Furthermore, we assume

$$\text{in}_w V = V(\text{in}_w f_1, \ldots, \text{in}_w f_k)$$

for all $w \in \mathbb{R}^n$. Then the equality of tropical cycles

$$\mathrm{Trop}(V) = \mathrm{trop}\, f_1 \cdots \mathrm{trop}\, f_k \cdot \mathbb{R}^r$$

holds.

Proof. The general idea (repeatedly used) is to show that we can localize the computations on both sides. Let Σ be a subdivision of $\mathbb{R}^r$ which is fine enough to compute $\mathrm{Trop}(V)$. In particular, assume that the rational functions $\mathrm{trop}\, f_i$ are affine on the cells of Σ, which also implies $\mathrm{in}_w(f_i) = \mathrm{in}_{w'}(f_i)$ for all $w, w' \in \mathrm{RelInt}(\tau), \tau \in \Sigma$. By formula 1.10, this means that every $w \in \mathrm{RelInt}(\tau)$, the polynomial $\mathrm{in}_w(f_i)$ is invariant under the action of the subtorus $T_\tau \subseteq T$. In other words, there exists a Laurent monomial x^{I_i} such that $f_i' := x^{I_i} \cdot \mathrm{in}_w(f_i) \in \kappa[V_\tau^\perp]$ and we can write

$$V(\mathrm{in}_w f_1, \ldots, \mathrm{in}_w f_k)/T_\tau = V(f_1', \ldots, f_k')$$
$$\subseteq T/T_\tau = \mathrm{Hom}(V_\tau^\perp, \kappa^*) = \mathrm{Spec}(\kappa[V_\tau^\perp]).$$

But note that these operations precisely correspond to computing the intersection product of the right hand side locally around τ. Namely, $\mathrm{trop}\,\mathrm{in}_w f_i$ is just the local part of $\mathrm{trop}\, f_i$ on $U(\tau)$ translated to the origin and multiplying with x^{I_i} just corresponds to adding a linear form such that the rational function vanishes on V_τ. Hence $\mathrm{trop}(x^{I_i} \cdot \mathrm{in}_w f_i)$ is a germ of $\mathrm{trop}\, f_i$ at τ. Let us now apply this locality for both sides to the case $\dim(\tau) = d$. By our assumption, the weight of τ in $\mathrm{Trop}(V)$ can be computed as the length of

$$V_\tau \cong \mathrm{in}_w V/T_\tau = V(f_1', \ldots, f_k'),$$

whereas the weight of τ on the right hand side can be computed by locality of the intersection product as the degree of

$$\operatorname{trop}(f'_1) \cdots \operatorname{trop}(f'_k) \cdot \mathbb{R}^r / V_\tau.$$

Thus applying theorem 1.6.1 to the system $f'_1, \ldots, f'_k$ finishes the proof.

□

Example 1.6.4
Let us give one simple example in the case when the genericity condition is not satisfied. Let $f = 2x+y+1-t$ and $g = x+y+1$ be two polynomials in two variables. Then the intersection $V = V(f, g)$ just consists of the point $(t, -1-t)$ whose image under valuation is $(1, 0)$. On the other hand, tropicalizing the polynomials we get

$$\operatorname{trop} f = \operatorname{trop} g = -\min\{x, y, 0\}$$

and therefore $\operatorname{trop} f \cdot \operatorname{trop} g \cdot \mathbb{R}^2 = \{(0, 0)\}$ with weight one. Thus the two processes do not coincide here. Things go wrong because

$$\operatorname{in}_{(1,0)} f = \operatorname{in}_{(1,0)} g = y + 1.$$

One might as well construct examples where points are completely missing or occur with different positive weights.

Remark 1.6.5
The theorem has a special consequence in the case where we consider only one polynomial $f \in \kappa[x_1^\pm, \ldots, x_r^\pm]$. Then the assumption is trivially satisfied and we conclude (together with previous remarks) that for any smooth toric variety whose fan refines the dual fan of $\operatorname{NP}(f)$ the Cartier divisor given by intersecting with the class of $V(f)$ and the Cartier divisor associated to $\operatorname{NP}(f)$ are the same. (For example, in [Fu93], the first lemma in section 5.5 could be formulated slightly stronger: The first inequality

in the first equation of the proof is in fact an equality.)

2 Tropical gravitational descendants

Introduction

This chapter is devoted to the study of tropical gravitational descendants. As in classical algebraic geometry, these are numbers obtained as the degrees of top-dimensional intersection products on the moduli space of parameterized tropical curves $\mathcal{M}_n^{\mathrm{lab}}(\mathbb{R}^r, \Delta)$ (classically, the moduli space of stable maps $\overline{M}_{0,n}(\mathbf{X}, \beta)$). The factors in these intersection products are pull backs along the evaluation morphisms on the one hand, and Psi-classes (resp. Psi-divisors, cf. 2.1.3) on the other hand. In tropical geometry, these numbers have a simple interpretation: We show that they are equal to the count of tropical curves which meet the pulled back incidence conditions in $\mathbb{R}^r$ and whose vertices have a prescribed valence (i.e. number of adjacent edges), due to the Psi-factors appearing in the intersection product. Hence, tropical gravitational descendants can basically be computed by combinatorial means, even if the combinatorics can be quite complicated in general.

But then, the next step is to relate the tropical numbers to the classical ones, which a priori do not admit such a combinatorial computation. Instead, the main tools to compute these numbers classically are the so-called WDVV equations and another set of equations, called topological recursion relations. These are recursive equations reflecting the recursive structure of the "boundary" of $\overline{M}_{0,n}(\mathbf{X}, \beta)$ — here "boundary" means the locus of reducible stable maps. Indeed, an irreducible component of this boundary is (nearly) the product of two smaller moduli spaces, where smaller means lower degree and fewer marked points.

The main content of the following text addresses the question if this recursive boundary structure is also apparent in the tropical moduli spaces (for rational curves) and if therefore tropical WDVV and topological recursion equations can be derived. It turns out that under a certain list of assumptions, this is possible (cf. theorems 2.4.5 and 2.4.8). Furthermore, we prove tropical versions of the well-known string, dilaton and divisor equations which deal with special cases of gravitational descendants.

As an application of this, we show in theorem 2.4.20 that certain tropical and classical gravitational descendants for rational curves in $\mathbb{P}^2$ and $\mathbb{P}^1 \times \mathbb{P}^1$ coincide — just because they satisfy the same set of equations which suffices to determine them from some initial values. Hence, the computation of the classical gravitational descendants in question can be reduced to the count of certain tropical curves with multiplicities. This extends the result of Mikhalkin in his fundamental work [Mi03], where he shows that the usual classical plane Gromov-Witten invariants (products without Psi-classes) can be computed by counting certain tropical curves as well (cf. theorem 1 in section 7 of [Mi03]).

However, it is important to keep in mind that the methods developed here are neither restricted to the plane nor (a priori) to specific toric varieties. Instead, to a large extent they are applicable to higher dimensions as well. For example, there is work in preparation by Andreas Gathmann and Eva-Maria Zimmermann which shows that the theory presented here can be used to obtain the same statement for rational Gromov-Witten invariants in $\mathbb{P}^r$, r arbitrary (cf. [GZ]). So there is good reason to hope that the equations proven here will be useful in many other applications.

This chapter essentially covers the material presented in [R08] and

[MR08]. In particular, the article [MR08] is joint work with Hannah Markwig. However, the presented approach is similar to [R08] and is to a large extent due to my own work. Particular contributions from Hannah Markwig are not presented here (e.g. the lattice path algorithm of [MR08, section 9]). Moreover, the proofs in subsection 2.4.3 are partially adapted from [GM05, section 5].

2.1 The moduli space of (abstract) rational tropical curves

In this section, we recall the construction of the moduli space of (abstract) rational tropical curves as a tropical variety and show first properties concerning the behaviour of "boundary and Psi-divisors". In particular, we recover a family property of the forgetful morphism ft_0 and the well-known dilaton and string equations concerning zero-dimensional products of Psi-divisors.

2.1.1 Smooth curves

Let us start with the definition of smooth curves. As the local model of a smooth curve, we will use the curve $L^r := L_1^r$ (cf. examples 1.1.1, 1.1.4 and 1.1.11).

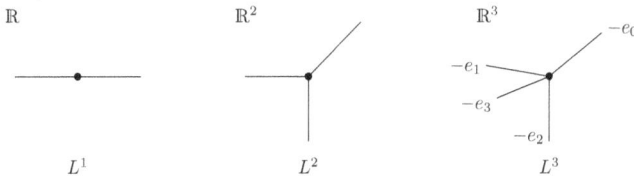

Note that L^r is irreducible for all r.

Let us make a small remark concerning our way of speaking. When we speak about tropical curves in the following, we mean, of course, one-dimensional tropical cycles. The word "abstract" indicates that the curves are considered modulo isomorphisms and is used in contrast to "parameterized" (curves together with a map into some vector space). Note also that for one-dimensional tropical cycles a "minimal" polyhedral structure exists (just remove all 2-valent vertices), which we will denote by the same letter as the cycle. The valence of a vertex ν, i.e. the number of edges containing ν, is denoted by $\mathrm{val}(\nu)$. Corresponding to the minimal polyhedral structure, our way of speaking is that a vertex of a curve is always at least 3-valent. But note that, as in the following definition, in most cases nothing changes if we regard every point of the curve as a possibly 2-valent vertex.

Definition 2.1.1 (Smooth curves)

A *smooth (abstract) curve* C is a one-dimensional connected tropical cycle that is locally isomorphic to L^r for suitable r, i.e. for each vertex ν in C there exists an isomorphism of tropical cycles $\mathrm{Star}_C(V) \cong L^{\mathrm{val}(V)}$. The *genus of C* is the first Betti number of $|C|$. Curves with genus 0 are called *rational curves*. In the following, we will only work with rational curves. An *n-marked smooth abstract curve* $(C, x_1, \ldots, x_n)$ is a smooth (abstract) curve C with exactly n unbounded rays (called *leaves*), which are labelled by $x_1, \ldots x_n$. If we instead label the leaves by elements of some finite set I, we will call it an *I-marked curve*.

Remark 2.1.2 (Smooth curves vs. metric graphs)

As no other abstract curves will be considered, we will often omit the word "smooth". Note that by definition a smooth curve is locally irreducible (cf. definition 1.2.27). As mentioned above, in the following isomorphic curves are identified (in the labelled case an isomorphism has to respect the labels, of course). In this sense, a smooth curve is in fact uniquely determined by its underlying metric graph (which is essentially the definition of an abstract curve in existing literature, in particular in [GKM07]), as the smoothness requirement fixes the tropical structure of such a metric graph uniquely. However, our "new" definition has the advantage that the balancing condition is incorporated in the curve C and need not, when using the "old" definition, be added to the definition of morphisms $C \to V$ later on (see [GKM07, definition 4.1]). Moreover, it is easy to check that any connected rational metric graph (without 1-valent vertices, but with unbounded edges) comes from a smooth tropical curve in our sense (inside some big vector space W). In particular, we will see this when analyzing the family property of ft_0 (cf. proposition 2.1.21).

Remark 2.1.3 (Smoothness criterion)
Let $\mathcal{F}$ be a one-dimensional fan in $V = \Lambda \otimes \mathbb{R}$ with $r + 1$ rays, all with weight 1 and generated by the primitive vectors $v_0, \ldots, v_r$. Let $V_{\mathcal{F}} := \mathbb{R}v_0 + \ldots + \mathbb{R}v_r$ be the vector space spanned by $\mathcal{F}$. It will be useful in the following to have a criterion to decide if $\mathcal{F}$ is smooth or not. The following equivalent conditions can be checked easily:

(a) $\mathcal{F}$ is isomorphic to L^r.

(b) The equations $v_0 + \ldots + v_r = 0$, $\dim(V_{\mathcal{F}}) = r$ and $V_{\mathcal{F}} \cap \Lambda = \mathbb{Z}v_0 + \ldots + \mathbb{Z}v_r$ hold.

(c) For arbitrary coefficients $\lambda_0, \ldots, \lambda_r \in \mathbb{R}$ the following equivalences hold:

   i) $\sum_{i=0}^{r} \lambda_i v_i = 0$ $\Leftrightarrow$ $\lambda_0 = \ldots = \lambda_r$ $\Leftrightarrow$ $\lambda_i - \lambda_j = 0$ for all i, j,

   ii) $\sum_{i=0}^{r} \lambda_i v_i \in \Lambda$ $\Leftrightarrow$ $\lambda_i - \lambda_j \in \mathbb{Z}$ for all i, j.

2.1.2 The tropical moduli space

Let us start with a brief motivation for the following definitions. In [Kap93], Kapranov constructed the moduli space of stable n-pointed genus zero curves $\overline{M}_{0,n}$ as the Chow resp. Hilbert quotient of the Grassmannian $G(2, n)$ parameterizing lines in $\mathbb{P}^{n-1}$ divided by the action of the torus

$T^{n-1} = T^n/T$ which dilates the coordinates of $\mathbb{P}^{n-1}$. This is based on the (older) observation that T^{n-1} acts freely on the open subset $G^0(2, n) \subseteq G(2, n)$ of lines with non-vanishing Plücker coordinates and that the respective quotient parameterizes configurations of n points in $\mathbb{P}^1$ modulo automorphisms, which equals $M_{0,n}$ (cf. [Kap93, introduction]). Thus there is a natural plan how to tropicalize $\overline{M}_{0,n}$ resp. $M_{0,n}$: Firstly, one tropicalizes the Plücker embedding of the Grassmannian in the sense of section 1.6, and secondly divides by the space of lineality of the tropicalization induced by the action of T^{n-1}. This was carried out (at least set-theoretically) in [SS04], and it turned out that the tropicalization of the Plücker embedding equals the so-called space of phylogenetic trees. Namely, if we identify a metric on the set $[n] := \{1, \ldots, n\}$ with a point in $\mathbb{R}^{\binom{n}{2}}$, then the set of metrics coming from n-marked metric trees (with possibly negative lengths on the leaves) forms a fan in $\mathbb{R}^{\binom{n}{2}}$ whose cones are in bijection with combinatorial types of trees and whose space of lineality is given by "star metrics" (metrics obtained from trees without inner edges; instead all leaves are adjacent to one single vertex).

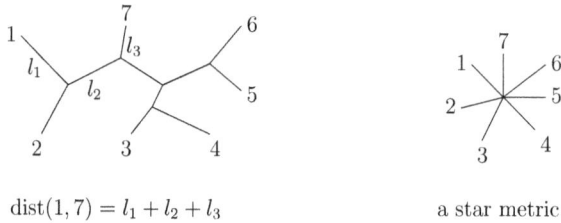

$$\operatorname{dist}(1, 7) = l_1 + l_2 + l_3 \qquad \text{a star metric}$$

Hence the quotient, denoted by $\mathcal{M}_n$, parameterizes metric trees with positive lengths on the bounded edges and the lengths of the leaves are forgotten/set to be zero/set to be infinite. It is known that $\mathcal{M}_n$ is a unimodular fan and that the closure of $M_{0,n}$ in the toric variety $\mathbf{X}(\mathcal{M}_n)$ equals $\overline{M}_{0,n}$ (cf. [Te04], [GM07], in both references it is theorem 5.5!). Moreover, in [GM07, theorem 1.2] the authors describe the equations of $\overline{M}_{0,n}$ in the Cox ring of $\mathbf{X}(\mathcal{M}_n)$ explicitly. In [GKM07, section 3], it is shown explicitly that $\mathcal{M}_n$ satisfies the balancing condition. We use the notation Φ_n from [GKM07] for the map

$$\Phi_n : \mathbb{R}^n \to \mathbb{R}^{\binom{n}{2}}$$
$$(a_1, \ldots, a_n) \mapsto ((a_i + a_j))_{ij},$$

whose image is the space of star metrics mentioned above. Note that this is indeed the tropicalization of the action of the torus T^n/T by dilating the coordinates, which is given on the Plücker embedding by the torus embedding

$$T^n/T \to T^{\binom{n}{2}}/T$$
$$(t_1, \ldots, t_n) \mapsto ((t_i \cdot t_j))_{ij},$$

where $T^{\binom{n}{2}}/T$ is the dense torus in $\mathbb{P}^{\binom{n}{2}-1}$.

Let us collect these definitions and some notations in the following:

Definition 2.1.4 (The moduli space of abstract curves)
The *moduli space of (abstract smooth) rational tropical curves*, denoted by $\mathcal{M}_n$ is the fan in $\mathbb{R}^{\binom{n}{2}}/\mathrm{Im}(\Phi_n)$ that parameterizes metric trees with positive lengths on the bounded edges and infinite lengths on the unbounded edges. Explicit descriptions can be found in [SS04, section 4], [GM07, section 5.2] and [GKM07, section 3]. The cones of $\mathcal{M}_n$ are in one-to-one correspondence with combinatorial types of n-marked trees (with 2-valent vertices removed), and the dimension of a cone equals the number of bounded edges in the respective combinatorial type. A general point in $\mathcal{M}_n$ (i.e. an element in the interior of a facet) is a 3-valent metric tree with $n - 3$ bounded edges (hence $\dim(\mathcal{M}_n) = n - 3$). All facets are equipped with weight 1, and $\mathcal{M}_n$ fulfills the balancing condition. By abuse of notation, we also denote the associated tropical cycle by $\mathcal{M}_n$. If we work with $\mathcal{M}_{n+1}$, the extra leaf is labelled by x_0. As $\mathcal{M}_3$ is just a

single point, we assume $n \geq 4$ in most cases. The notation $I|J$ denotes a non-trivial partition of $[n] = \{1, \ldots, n\}$ (or of $\{0\} \cup [n]$ if we work with $\mathcal{M}_{n+1}$) into the two disjoint subsets I and J. In most cases — the few exceptions will be mentioned — we will consider this partition to be unordered. Occasionally, we use I^c to denote the complement of I and write $I|I^c$. If $|I| \neq 1 \neq |J|$, such a partition describes a ray in $\mathcal{M}_n$ generated by the metric tree $V_{I|J} \in \mathcal{M}_n$ with only one bounded edge:

$$V_{I|J} := \begin{array}{c} x_i, \\ i \in I \end{array} \!\!\!\! \searrow\!\!\!\!\!\!\!\!\!\!\!\!\!\!\nearrow \begin{array}{c} x_j, \\ j \in J \end{array} \quad \in \mathcal{M}_n.$$
$$\underset{\text{edge of length 1}}{\uparrow}$$

An edge of a tree is uniquely determined by the partition $I|J$ of the leaves if we remove the edge. In this sense, we can regard the partitions $I|J$ as "global" labels of the edges of a tree, where $I|J$ labels the leaf x_i if $I = \{i\}$ or $J = \{i\}$, and a bounded edge otherwise. A cone τ of $\mathcal{M}_n$ is generated by the vectors $V_{I|J}$ for all partitions which correspond to edges in the combinatorial type of τ. In particular, it is natural to use the lengths of the bounded edges as local coordinates of a cone of $\mathcal{M}_n$ — this identifies each cone τ of $\mathcal{M}_n$ with the positive orthant of $\mathbb{R}^{\dim(\tau)}$.

Remark 2.1.5

Let us make some remarks here.

- We sometimes also think of $V_{I|J}$ as a vector in $\mathbb{R}^{\binom{n}{2}}$, in which case we also allow $|I| = 1$ or $|J| = 1$ to get simpler formulas. However, as $V_{\{k\}|[n]\setminus\{k\}} = \Phi_n(0, \ldots, 0, 1, 0, \ldots, 0)$, these vectors vanish modulo $\mathrm{Im}(\Phi_n)$.

- Note that for the following purposes, the underlying lattice of $\mathbb{R}^{\binom{n}{2}}/\Phi_n(\mathbb{R}^n)$ is *not* $\mathbb{Z}^{\binom{n}{2}}/\Phi_n(\mathbb{Z}^n)$, but is the lattice generated by the vectors $V_{I|J}$, denoted by Λ_n (see [GKM07, 3.3]). This is a technical issue, as it does not change the lattices of the cones $\Lambda_\tau, \tau \in \mathcal{M}_n$, but is necessary to make maps such as forgetful maps *integer* affine.

- We prove explicitly in proposition 2.1.21 that $\mathcal{M}_n$ really parameterizes smooth rational tropical curves in the sense of definition 2.1.1. Therefore we often speak of (smooth abstract) curves instead of metric trees in the following — without different meaning.

- The "recursive structure" of $\mathcal{M}_n$ can be observed by analyzing its star fans around cells. Let τ be a cone of $\mathcal{M}_n$ and let Γ denote the corresponding combinatorial type of n-marked trees. Then it is easy to check that the star around τ satisfies

$$\text{Star}_{\mathcal{M}_n}(\tau) = \prod_{\substack{\nu \text{ vertex} \\ \text{of } \Gamma}} \mathcal{M}_{\text{val}(\nu)}.$$

Next, we define divisors respectively rational functions that play the role of "boundary" divisors in our moduli space. They all lie in the codimension one skeleton of $\mathcal{M}_n$, therefore represent higher-valent curves. Note that our nomenclature is a bit confusing here. Even if we call all curves parameterized by $\mathcal{M}_n$ smooth, we consider the codimension one skeleton of $\mathcal{M}_n$ as (part of) the boundary of $\mathcal{M}_n$ which classically consists of singular curves.

As $\mathcal{M}_n$ is unimodular, we can define rational functions on $\mathcal{M}_n$ as described in example 1.2.2 (e).

Definition 2.1.6 (Boundary functions)
Let $I|J$ be a partition with $|I| \neq 1 \neq |J|$. Then $\varphi_{I|J}$ denotes the unique rational function with

$$\varphi_{I|J}(V_{I'|J'}) := \begin{cases} 1 & \text{if } I = I' \text{ or } I = J', \\ 0 & \text{otherwise.} \end{cases}$$

If Γ is the metric tree represented by a point $p \in \mathcal{M}_n$, then $\varphi_{I|J}(p)$ is just the length of the bounded edge $I|J$ in Γ (being zero if no such edge exists). Furthermore, we use the notation

$$\varphi_{k,l} := \varphi_{\{k,l\}|[n]\setminus\{k,l\}}$$

for $k \neq l \in [n]$.

The ridges of $\mathcal{M}_n$ correspond to combinatorial types of curves with one 4-valent vertex, which we denote like this:

$$\begin{smallmatrix} A \\ D \end{smallmatrix} \times \begin{smallmatrix} B \\ C \end{smallmatrix}$$

Here A, B, C and D denote the four parts of the combinatorial type adjacent to the 4-valent vertex and by abuse of notation also the sets of leaves belonging to this part (as, in most cases, this is the only information needed).

In order to compute the weight of a ridge $\begin{smallmatrix} A \\ D \end{smallmatrix} \times \begin{smallmatrix} B \\ C \end{smallmatrix}$ in the divisor of a rational function on $\mathcal{M}_n$, let us have a look at $\mathrm{Star}_{\mathcal{M}_n}(\begin{smallmatrix} A \\ D \end{smallmatrix} \times \begin{smallmatrix} B \\ C \end{smallmatrix})$. In fact, it is easy to see that $\mathrm{Star}_{\mathcal{M}_n}(\begin{smallmatrix} A \\ D \end{smallmatrix} \times \begin{smallmatrix} B \\ C \end{smallmatrix})$ contains three facets corresponding to the three types of removing the 4-valent vertex by inserting a new bounded edge.

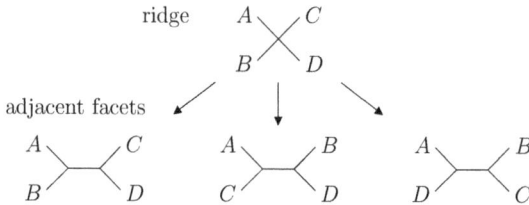

The primitive representatives are $V_{A \cup B | C \cup D}$, $V_{A \cup C | B \cup D}$ and $V_{A \cup D | B \cup C}$. For the balancing condition around $\begin{smallmatrix} A \\ D \end{smallmatrix} \times \begin{smallmatrix} B \\ C \end{smallmatrix}$, it suffices to show the equation

$$V_{A \cup B | C \cup D} + V_{A \cup C | B \cup D} + V_{A \cup D | B \cup C}$$
$$= V_{A | B \cup C \cup D} + V_{B | A \cup C \cup D} + V_{C | A \cup B \cup D} + V_{D | A \cup B \cup C},$$

as all vectors on the right hand side lie in the vector space spanned by the ridge $\frac{A}{D} \times \frac{B}{C}$, as required. But the equation follows from the fact that, on the level of metric trees, the distance between two marked leaves is identical on both sides: If both leaves belong to the same set A, B, C, D, the distance is 0, if not, it is 2.

This discussion also shows that $\mathcal{M}_n$ is locally irreducible (cf. example 1.2.28 (c); and, in particular, $\mathcal{M}_n$ is irreducible, cf. lemma 1.2.29). This implies that for any rational function φ on $\mathcal{M}_n$ the equation $|\operatorname{div}(\varphi)| = |\varphi|$ holds (cf. lemma 1.2.31).

Let us now compute the divisors of the functions $\varphi_{I|J}$. In the following, a formula involving I, J and A, B, C, D stands for all permuted formulas as well, e.g. $I = A$ means "$I = A$ or $I = B$ or $J = A$ ...".

Lemma 2.1.7 (Boundary divisors)
The boundary divisor $\operatorname{div}(\varphi_{I|J})$ *carries the weight function*

$$\omega_{\varphi_{I|J}}\left(\tfrac{A}{D} \times \tfrac{B}{C}\right) = \begin{cases} 1 & \text{if } I = A \cup B, \\ -1 & \text{if } I = A, \\ 0 & \text{otherwise.} \end{cases}$$

These divisors were computed before by Matthias Herold (cf. [H07]).

Proof. Following from the previous discussion, the weight of $\frac{A}{D} \times \frac{B}{C}$ in $\operatorname{div}(\varphi_{I|J})$ is by definition

$$\begin{aligned}
\omega_{\varphi_{I|J}}\left(\tfrac{A}{D} \times \tfrac{B}{C}\right) = \; & \varphi_{I|J}(V_{A \cup B | C \cup D}) + \varphi_{I|J}(V_{A \cup C | B \cup D}) + \varphi_{I|J}(V_{A \cup D | B \cup C}) \\
& - \varphi_{I|J}(V_{A | B \cup C \cup D}) - \varphi_{I|J}(V_{B | A \cup C \cup D}) \\
& - \varphi_{I|J}(V_{C | A \cup B \cup D}) - \varphi_{I|J}(V_{D | A \cup B \cup C}).
\end{aligned}$$

Hence, this weight is 1 if I is the union of two of the sets A, B, C, D and is -1 if I equals one of the four sets. Otherwise, it is 0. □

Remark 2.1.8

It is clear from remark 1.2.17 and the discussion at the beginning of this subsection (cf. also [GM05, lemma 5.10]) that the functions $\varphi_{I|J}$ respectively the divisors $\mathrm{div}(\varphi_{I|J})$ are the tropical analogues of the irreducible components of the boundary of the moduli space of stable curves $\overline{M}_{0,n}$. This is justified by the fact that in the following we will re-prove many of the classical algebro-geometric statements concerning the intersection-theoretic behaviour of boundary divisors.

However, note that we will not use this relationship to the classical world explicitly. Instead, as preparation for the upcoming section on parameterized curves, we choose a different approach and prove the following statements purely "inside tropical geometry", i.e. by combinatorial arguments.

As mentioned, many of the following statements are inspired by analogue statements in the algebro-geometric theory of Gromov-Witten invariants. The only reference I could find that contains all occurring statements in a similar notation is an unpublished, but available online paper "Notes on psi classes" [Ko] by Joachim Kock. Hence, for the sake of comparing the assertions easily, the following statements include [Ko] references indicating the place where the reader can find the corresponding classical statement.

Lemma 2.1.9 (cf. [Ko] 1.2.5)

The equation

$$\varphi_{i,j} \cdot \varphi_{i,k} \cdot \mathcal{M}_n = 0$$

holds for $n \geq 4$ and pairwise different $i, j, k \in [n]$.

Proof. By lemma 2.1.7 we know that for each metric tree corresponding to a point in $|\varphi_{i,k}| = |\mathrm{div}(\varphi_{i,k})|$, the leaves x_i and x_k are adjacent to the same vertex. In particular, such a metric tree cannot contain an edge separating $\{i,j\}|\{i,j\}^c$. But $\varphi_{i,j}$ just measures the length of such an edge if present. Hence $\varphi_{i,j}$ vanishes on $|\varphi_{i,k}|$, which proves the claim. $\qquad\square$

2.1.3 Psi-divisors

Analogues of Psi-classes on tropical $\mathcal{M}_n$ have been defined recently by G. Mikhalkin ([Mi07]). These objects are Weil divisors in $\mathcal{M}_n$ a priori. However, as we want to perform intersections with them, we need Cartier divisors/rational functions that describe these Weil divisors. Let us make a remark concerning this approach first.

Remark 2.1.10 (Complete intersections)
Let Z be an arbitrary ambient cycle, not necessarily $\mathbb{R}^r$ (e.g. $\mathcal{M}_n$). A subcycle X of Z is called a *complete intersection in Z* if it can be obtained as an intersection product $X = \varphi_1 \cdots \varphi_l \cdot Z$ for suitable Cartier divisors φ_i on Z (where $l = \dim(X) - \dim(Z)$).

Let X, Y be two complete intersections in Z, given by $X = \varphi_1 \cdots \varphi_l \cdot Z$ and $Y = \varphi'_1 \cdots \varphi'_{l'} \cdot Z$. Then we define their intersection product

$$X \cdot Y := \varphi_1 \cdots \varphi_l \cdot \varphi'_1 \cdots \varphi'_{l'} \cdot Z.$$

Using the commutativity of the intersection product of Cartier divisors this multiplication is independent of the chosen functions, commutative and satisfies $|X \cdot Y| \subseteq |X| \cap |Y|$. Moreover, if $X = V = \Lambda \otimes \mathbb{R}$, it follows from corollary 1.5.6 that this definition coincides with the usual intersection product of cycles.

We also would like to pull back a complete intersection $X = \varphi_1 \cdots \varphi_l \cdot Z$ along a morphism $f : Z' \to Z$ defining

$$f^*(X) := f^*(\varphi_1) \cdots f^*(\varphi_l) \cdot Z'.$$

However, in general this definition is not independent of the chosen functions $\varphi_1, \ldots, \varphi_l$ (e.g. pull back the rational function from remark 1.2.7 (5. item) to one of the coordinate axes). But it works in the case when $f : Z \times Z' \to Z$ is a projection, as in this case lemma 1.5.4 guarantees that $f^*(X) = X \times Z'$ holds.

More general, these definitions and statements hold for sums of complete intersections as well.

Our next step is to introduce "Psi-classes" as Weil divisors of rational functions on $\mathcal{M}_n$. We use the notion "Psi-divisor" instead of "Psi-class" to emphasize that, in contrast to the algebro-geometric case, tropically Psi-divisors are not defined up to rational equivalence.

The definition of Psi-divisors as Weil divisors of rational functions and their intersections were first studied in [KM07]. Let us recall the important definitions and results of [KM07] here.

Definition 2.1.11 (Psi-functions)
For $k \in [n]$, we define the *k-th Psi-function* ψ_k to be the unique rational function on $\mathcal{M}_n$ with

$$\psi_k(V_{I|J}) := \frac{|I|(|I| - 1)}{(n-1)(n-2)}$$

for all partitions $I|J$ with $|I|, |J| \geq 2$ and $k \in J$.

Remark 2.1.12
Our function ψ_k equals the function $\frac{1}{\binom{n-1}{2}} f_k$ defined in [KM07] (follows from [KM07, lemma 2.6]). In particular, ψ_k is a convex function (cf. [KM07, remark 2.5]).

Obviously the numbers $\psi_k(V_{I|J})$ are only rational. A generalization of intersection theory to rational numbers is straightforward, but nearly unnecessary: The weights of the divisor of ψ_k turn out to be integers (see the following proposition) and there exist integer rational functions providing the same divisor (see lemma 2.1.26). This particular function

ψ_k was chosen in [KM07] because of its symmetry.

Proposition 2.1.13 (Psi-divisors, see [KM07] 3.5)
The k-th Psi-divisor $\operatorname{div}(\psi_k)$ consists of the ridges $_D^A\times_C^B$ corresponding to trees where the marked leaf x_k lies at the 4-valent vertex, i.e. the weight function of $\operatorname{div}(\psi_k)$ is

$$\omega_{\psi_k}(_D^A\times_C^B) = \begin{cases} 1 & \text{if } \{k\} = A, \\ 0 & \text{otherwise.} \end{cases}$$

Remark 2.1.14
In [Mi07, definition 3.1], Mikhalkin suggests that $\operatorname{div}(\psi_k)$ is the correct tropicalization of the k-th Psi-class on $\overline{M}_{0,n}$. His motivation for this definition is a direct translation of the classical definition (pull back of the cotangent bundle of the universal family along the k-th section). Another motivation is the following: As explained in subsection 1.1.4, the tropicalization of a cohomology class γ is given by the tropical cycle whose weight on a cone τ is the degree of the intersection $\gamma \cap [V(\tau)]$. In our case, $\overline{M}_{0,n}$ is embedded as a subvariety in the toric variety $\mathbf{X}(\mathcal{M}_n)$ whose orbit structure corresponds to the stratification of $\overline{M}_{0,n}$ in types of reducible curves. Hence a ridge $_D^A\times_C^B$ determines a one-dimensional subvariety $D \subseteq \overline{M}_{0,n}$ of stable n-marked rational curves whose dual graph corresponds to the combinatorial type of $_D^A\times_C^B$. More precisely, D is isomorphic to $\overline{M}_{0,4} \cong \mathbb{P}^1$ and consists of curves with several components with 3 special points and precisely one component with 4 special points.

This suggests that the weight of the tropicalization of a classical Psi-class Ψ_k on the ridge $_D^A\times_C^B$ is equal to the degree of the intersection of $\Psi_k \cap [D]$. But this can be computed explicitly. If none of the 4 special points on the distinguished component is the marked point x_k, then the restriction of Ψ_k to D is trivial, i.e. $\deg(\Psi_k \cap [D]) = 0$. If $A = \{k\}$, i.e. if one of the four special points is the marked point x_k, then the restriction of ψ_k to D just corresponds to the respective Psi-class on $\overline{M}_{0,4}$, therefore $\deg(\Psi_k \cap [D]) = 1$. So this construction provides precisely the Weil divisor $\operatorname{div}(\psi_k)$ in $\mathcal{M}_n$, our tropical Psi-divisor. (see also [Ka09, section 7] for a more detailed and general treatment).

Note that, a priori, we are only interested in the divisor $\operatorname{div}(\psi_k)$ and not in the describing function ψ_k. The choice of ψ_k is only necessary to compute intersections with the Psi-divisors. Note also that, as long as we only intersect complete intersections of $\mathcal{M}_n$ (which we always do), the choice of the particular function describing $\operatorname{div}(\psi_k)$ does not matter (cf. remark 2.1.10).

Notation 2.1.15 (The tau-notation)
Adapted from the notation often used in the classical situation, we will introduce the following τ-notation that makes formulas shorter and hides "unimportant" data such as the number of marked leaves. For integers $a_1, \ldots, a_n$ we define

$$(\tau_{a_1} \cdots \tau_{a_n}) := \psi_1^{a_1} \cdots \psi_n^{a_n} \cdot \mathcal{M}_n$$

if all integers are non-negative, otherwise $(\tau_{a_1} \cdots \tau_{a_n}) = \emptyset$. Every factor τ_{a_k} stands for a marked leaf and the index a_k serves as the exponent with which the corresponding Psi-function appears in the intersection product. If $\sum a_k = \dim(\mathcal{M}_n) = n - 3$, the above cycle is zero-dimensional (in fact, its only point corresponds to the curve without bounded edges where all leaves are adjacent to one single vertex) and we define

$$\langle \tau_{a_1} \cdots \tau_{a_n} \rangle := \deg \left(\psi_1^{a_1} \cdots \psi_n^{a_n} \cdot \mathcal{M}_n \right).$$

The main theorem of [KM07] computes these intersection products of Psi-divisors:

Theorem 2.1.16 (Products of Psi-divisors, see [KM07] 4.1)
The intersection product $(\tau_{a_1} \cdots \tau_{a_n})$ is the subfan of $\mathcal{M}_n$ consisting of the closure of the cones of dimension $n - 3 - \sum_{i=1}^{n} a_i$ whose interior curves C have the following property:

Let $k_1, \ldots, k_q \in N$ be the marked leaves adjacent to a vertex ν of C. Then the valence of V is

$$\mathrm{val}(V) = a_{k_1} + \ldots + a_{k_q} + 3.$$

Let us define the multiplicity of this vertex to be $\mathrm{mult}(\nu) := \binom{\mathrm{val}(\nu)-3}{a_{k_1},\ldots,a_{k_q}}$. Then the weight of such a cone σ in X is

$$\omega_X(\sigma) = \prod_{\nu} \mathrm{mult}(\nu),$$

where the product runs through all vertices ν of an interior curve of σ.

In this section we re-prove the zero-dimensional case of this theorem (see remark 2.1.24). To do this, we first have to analyze how Psi- and boundary divisors intersect and how they behave when pulled back or pushed forward along forgetful morphisms.

Lemma 2.1.17 (cf. [Ko] 1.2.7)
It holds that

$$\varphi_{i,j} \cdot \psi_i \cdot \mathcal{M}_n = 0$$

for $n \geq 4$ and $i \neq j \in [n]$.

Proof. Curves in $|\psi_i|$ cannot contain a bounded edge with partition $\{i,j\}|\{i,j\}^c$, as the leaf x_i does not lie at a 3-valent vertex. Thus $\varphi_{i,j}$ vanishes on $|\psi_i|$. $\qquad\square$

2.1.4 The forgetful morphism

The forgetful map $\mathcal{M}_{n+1} \to \mathcal{M}_n$ that forgets the extra leaf x_0 (and then removes the possibly occurring 2-valent vertex) is denoted by ft_0 (cf. [GM05, definition 4.1] and [GKM07, definition 3.8]). By [GKM07, proposition 3.9] this map is a tropical morphism. Note that the image of a cone of $\mathcal{M}_{n+1}$ under ft_0 is a cone in $\mathcal{M}_n$, therefore no refinement is necessary to compute push forwards in the following. Note also that a cone is mapped injectively if and only if x_0 is adjacent to a higher-valent vertex in the corresponding combinatorial type of trees (as otherwise, after removing the 2-valent vertex, there is one bounded edge less). Let us first pull back Psi-functions along ft_0.

Proposition 2.1.18 (Pull backs of Psi-functions, cf. [Ko] 1.3.1)
Let $n \geq 4$ and let $\mathrm{ft}_0 : \mathcal{M}_{n+1} \to \mathcal{M}_n$ be the morphism that forgets the leaf x_0. For $k \in [n]$ it holds that

$$\mathrm{div}(\psi_k) = \mathrm{div}(\mathrm{ft}_0^* \, \psi_k) + \mathrm{div}(\varphi_{0,k}).$$

Proof. This can be proven by explicitly computing the weights of the three divisors. We distinguish four cases (up to renaming A, B, C and D):

$\omega_f\left(\begin{smallmatrix}A\\D\end{smallmatrix}\times\begin{smallmatrix}B\\C\end{smallmatrix}\right)$	$f = \psi_k$	$f = \mathrm{ft}_0^* \, \psi_k$	$f = \varphi_{0,k}$
$A = \{0, k\}$	0	1	-1
$A = \{0\}, B = \{k\}$	1	0	1
$A \supsetneq \{0\}, B = \{k\}$	1	1	0
otherwise	0	0	0

$\square$

To keep formulas shorter, we omit "$\cdot \mathcal{M}_n$" when it is clear from the context that we denote an intersection product on $\mathcal{M}_n$.

Corollary 2.1.19 (cf. [Ko] 1.3.2 and 1.3.3)
Let $n \geq 4$ and let $\mathrm{ft}_0 : \mathcal{M}_{n+1} \to \mathcal{M}_n$ be the morphism that forgets the leaf x_0. Then for $k \in [n]$ the following formulas hold:

(a)
$$\varphi_{0,k}^2 = -\,\mathrm{ft}_0^*(\psi_k) \cdot \varphi_{0,k}$$

(b)
$$\psi_k^a = \mathrm{ft}_0^*(\psi_k)^a + \mathrm{ft}_0^*(\psi_k)^{a-1} \cdot \varphi_{0,k}$$

(c)
$$\psi_k^a = \mathrm{ft}_0^*(\psi_k)^a + (-1)^{a-1}\varphi_{0,k}^a$$

Proof. All the formulas are easy applications of lemmata 2.1.17 and 2.1.18. $\square$

Lemma 2.1.20
Let $n \geq 4$, let $\mathrm{ft}_0 : \mathcal{M}_{n+1} \to \mathcal{M}_n$ be the morphism that forgets the leaf x_0 and choose $k \in [n]$. Then

$$\mathrm{ft}_{0*}(\mathrm{div}(\varphi_{0,k})) = \mathrm{ft}_{0*}(\mathrm{div}(\psi_k)) = \mathcal{M}_n.$$

Proof. We show that $\mathrm{ft}_{0*}(\mathrm{div}(\varphi_{0,k})) = \mathcal{M}_n$ by direct computation: Let σ' be a facet of $\mathcal{M}_n$ corresponding to a 3-valent combinatorial type. Let ν be the vertex adjacent to x_k. Then there exists precisely one cone σ in $\mathrm{div}(\varphi_{0,k})$ whose image under ft_0 is σ', namely the cone obtained by attaching the additional leaf x_0 to the vertex ν. Moreover, on such a cone, the length of the bounded edges remain unchanged under ft_0 and therefore $\mathrm{ft}_0(\Lambda_\sigma) = \Lambda_{\sigma'}$. On the other hand, cones in $\mathrm{div}(\varphi_{0,k})$ with negative weight are not mapped injectively, as in this case x_0 is adjacent to a 3-valent vertex and stabilization is needed. This shows that $\mathrm{ft}_{0*}(\mathrm{div}(\varphi_{0,k})) = \mathcal{M}_n$. The equation $\mathrm{ft}_{0*}(\mathrm{div}(\psi_k)) = \mathcal{M}_n$ follows from the same argument or by using lemma 2.1.18, the projection formula and $\mathrm{ft}_{0*}(\mathcal{M}_{n+1}) = 0$ (because the dimension is too high). $\qquad\qquad\qquad\qquad\qquad\qquad\qquad\qquad\qquad\qquad\square$

It is well-known that for the classical moduli space $\overline{M}_{0,n}$, the forget-ful morphism plays the role of the universal family (cf. [KV07, section 1.3]). So far, in the tropical setting we can can only prove the following statement .

Proposition 2.1.21 (Family property of ft_0 for abstract curves)
Let p be a point in $\mathcal{M}_n$ and let $C_p = \mathrm{ft}_0^{-1}(p)$ be the fibre of p under the forgetful morphism $\mathrm{ft}_0 : \mathcal{M}_{n+1} \to \mathcal{M}_n$. Then the following holds:

(a) C_p has a canonical structure of a one-dimensional polyhedral complex.

(b) The leaves of C_p (as a graph itself) are the facets where x_0 and an-other leaf x_i lie at the same 3-valent vertex (i.e. the leaves are the sets $L_i := \overline{\{y \in C_p | \varphi_{0,i}(y) > 0\}}$). Moreover $p \in \mathcal{M}_n$ represents the n-marked metric graph $(C_p, L_1, \ldots, L_n)$.

(c) When we equip all its facets with weight 1, C_p is a smooth curve (in the sense of definition 2.1.1).

(d) Let $\sum_k \mu_k p_k = \varphi_1 \cdots \varphi_{n-3} \cdot \mathcal{M}_n$ be a zero-dimensional cycle in $\mathcal{M}_n$ obtained as the intersection product of convex functions φ_j. Then

$$\mathrm{ft}_0^*(\varphi_1) \cdots \mathrm{ft}_0^*(\varphi_{n-3}) \cdot \mathcal{M}_{n+1} = \sum_k \mu_k C_{p_k}.$$

We write this as $\mathrm{ft}_0^*(\sum_k \mu_k p_k) = \sum_k \mu_k C_{p_k}$.

Proof. (a): As polyhedral complex, C_p consists of the polyhedra $(\mathrm{ft}_0\,|_\sigma)^{-1}(p)$ $= C_p \cap \sigma$ for each cone σ of $\mathcal{M}_{n+1}$. The dimension of these polyhedra can be at most one as $\dim(\mathrm{ft}_0(\sigma)) \geq \dim(\sigma) - 1$ (it depends on whether x_0 is adjacent to a 3-valent or higher-valent vertex).

(b): Let Γ_p denote the n-marked metric graph represented by p. The bijective map $\Gamma_p \to C_p$ indicated in the picture identifies the two graphs.

$x_1 \quad \Gamma_p \quad x_5 \qquad x_4 \qquad \bullet \qquad \longmapsto \qquad x_1 \qquad x_5 \qquad x_4 \qquad \in C_p$
$x_2 \qquad x_3 \qquad\qquad x_2 \qquad x_3 \qquad x_0$

(c): Let ν be a vertex of C_p. It corresponds to the metric graph Γ_p with the extra leaf x_0 adjacent to one of the vertices. Let us label the other edges containing this vertex by $1, \ldots, m$ and let us divide the other leaves $[n] = I_1 \cup \ldots \cup I_m$ according to via which edge one gets from x_0 to x_i. There are m facets in C_p containing ν, corresponding to moving x_0 on one of the edges. Hereby one has to shorten the edge $I_k|I_k^c$ as much as the length of $I_k \cup \{x_0\}|(I_k \cup \{x_0\})^c$ increases.

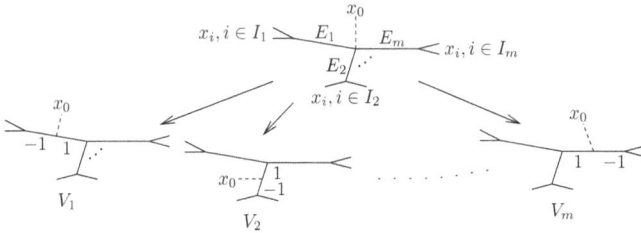

x_0
$x_i, i \in I_1 \quad E_1 \qquad E_m \quad x_i, i \in I_m$
E_2
$x_i, i \in I_2$
x_0
$-1 \quad 1 \qquad\qquad x_0 \cdots \frac{1}{-1} \qquad \cdots\cdots\cdots \qquad 1 \quad -1$
$V_1 \qquad\qquad V_2 \qquad\qquad V_m$
$x_0 \qquad\qquad\qquad\qquad\qquad x_0$

Thus the primitive integer vector of the corresponding edge with respect to ν is given by

$$V_k := V_{I_k \cup \{x_0\}} - V_{I_k},$$

where we use the shorthand $V_I := V_{I|I^c}$ in the following. Note that this formula as well as the following ones also hold in the case that I_k consists only of a single leaf x_i (which means x_i is adjacent to the same vertex as x_0), as $V_{\{x_i\}} = 0 \in \mathbb{R}^{\binom{n+1}{2}}/\mathrm{Im}(\Phi_{n+1})$. To prove the statement we now use criterion 2.1.3 (c) and verify the conditions i) and ii), which can be done by applying some formulas of [KM07]. Let $\mathcal{S}$ be the set of two-element subsets of $[n]$ (i.e. not containing 0). It follows from [KM07, 2.3, 2.4, 2.6] that the vectors $V_S, S \in \mathcal{S}$ fulfill i) and ii) (with $V = \mathbb{R}^{\binom{n+1}{2}}/\mathrm{Im}(\Phi_{n+1})$

and $\Lambda = \Lambda_n$). Furthermore [KM07, 2.6] gives us a representation of our vectors in terms of the vectors V_S, namely

$$V_{I_k} = \sum_{\substack{S \in \mathcal{S} \\ S \subseteq I_k}} V_S$$

$$V_{I_k \cup \{x_0\}} = \sum_{\substack{S \in \mathcal{S} \\ S \cap I_k = \emptyset}} V_S = -\left(\sum_{\substack{S \in \mathcal{S} \\ S \cap I_k \neq \emptyset}} V_S \right),$$

and therefore

$$V_k = -\left(\sum_{S \in \mathcal{S}} |S \cap I_k| \cdot V_S \right).$$

Now let $\lambda_1, \ldots, \lambda_m$ be arbitrary real coefficients. Then we obtain the formula

$$\sum_{k=1}^{m} \lambda_k V_k = -\left(\sum_{\substack{\{i,j\} \in \mathcal{S} \\ i \in I_k, j \in I_{k'}}} (\lambda_k + \lambda_{k'}) \cdot V_{\{i,j\}} \right).$$

Now all differences of two coefficients on the left hand side $\lambda_k - \lambda_k'$ can be obtained as differences of two coefficients on the right hand side (choose elements $i \in I_k, j \in I_{k'}, l \in I_{k''}$; then the coefficients of $V_{\{i,l\}}$ and $V_{\{j,l\}}$ differ by $\lambda_k + \lambda_{k''} - \lambda_{k'} - \lambda_{k''} = \lambda_k - \lambda_{k'}$). Conversely, a right hand side difference of coefficients equals the sum of two left hand side differences. (The coefficients of $V_{\{i_1,i_2\}}$ and $V_{\{j_1,j_2\}}$ differ by $(\lambda_{k_1} - \lambda_{l_1}) + (\lambda_{k_2} - \lambda_{l_2})$, where $i_1 \in I_{k_1}, i_2 \in I_{k_2}, j_1 \in I_{l_1}, j_2 \in I_{l_2}$.) Hence, as conditions 2.1.3 i) and ii) hold for the vectors V_S, they also hold for the vectors V_k. This proves that C_p (with trivial weights) forms a smooth curve.

(d): First of all, the set-theoretic equation

$$|\mathrm{ft}_0^*(\varphi_1) \cdots \mathrm{ft}_0^*(\varphi_{n-3}) \cdot \mathcal{M}_{n+1}| \subseteq \mathrm{ft}_0^{-1}(|\varphi_1 \cdots \varphi_{n-3} \cdot \mathcal{M}_n|) = \bigcup_k |C_{p_k}|.$$

follows by induction from lemma 1.2.25 (note that the pull back of a convex function is convex again). But the sets $|C_{p_k}|$ are pairwise disjoint (as they are fibres of pairwise different points) and belong to irreducible cycles (as the curves C_{p_k} are smooth abstract curves). Thus any one-dimensional cycle whose support lies in $\bigcup_i |C_{p_k}|$ is actually a sum $\sum_k \lambda_k C_{p_k}, \lambda_k \in \mathbb{Z}$. So it remains to check that in our case the coefficients λ_k coincide with μ_k. To do this, we choose an arbitrary leaf $x_i \neq x_0$ and consider the function $\varphi_{0,i}$ on C_{p_k}. On the leaf L_i of C_{p_k}, where x_0 and x_i are adjacent to the same 3-valent vertex, it measures the length of the third edge, elsewhere it is constantly zero. Thus $\varphi_{0,i} \cdot C_{p_k} = V_{p_k}$, where V_{p_k} is the vertex of C_{p_k} adjacent to L_i (where x_0 and x_i lie together at a higher-valent vertex). Thus we get

$$\mathrm{ft}_{0*}\left(\varphi_{0,i} \cdot \left(\sum_k \lambda_k C_{p_k}\right)\right) = \mathrm{ft}_{0*}\left(\sum_k \lambda_k V_{p_k}\right) = \sum_k \lambda_k p_k.$$

But then, we can use the projection formula and lemma 2.1.20 and compute

$$\mathrm{ft}_{0*}\left(\varphi_{0,i} \cdot \mathrm{ft}_0^*(\varphi_1) \cdots \mathrm{ft}_0^*(\varphi_{n-3}) \cdot \mathcal{M}_{n+1}\right)$$
$$= \varphi_1 \cdots \varphi_{n-3} \cdot \mathrm{ft}_{0*}(\varphi_{0,i} \cdot \mathcal{M}_{n+1}) = \sum_k \mu_k p_k.$$

Comparing the coefficients proves the statement. $\qquad\square$

Remark 2.1.22

Hence there is a bijection between metric trees and smooth rational curves according to definition 2.1.1 (modulo isomorphisms). In particular, $\mathcal{M}_n$ parameterizes smooth (abstract) rational tropical curves modulo isomorphism, and the curve associated to a point $p \in \mathcal{M}_n$ is given by the preimage $\mathrm{ft}_0^{-1}(p)$.

In fact, the statement about the fibers of ft_0 can be generalized: It is easy to see that the preimage of every tropical cycle X in $\mathcal{M}_n$ under ft_0 is again a tropical cycle (with canonical weights).

2.1.5 First equations and numbers

We now prove the tropical analogues of the well-known classical equations concerning Psi-products such as the string and dilaton equation. Of course, the idea behind this is to extend these equations to the case of parameterized curves later on.

Theorem 2.1.23 (String equation for abstract curves, cf. [Ko] 1.4.2)

For zero-dimensional intersection products of Psi-divisors the following holds:

$$\langle \tau_0 \prod_{k=1}^{n} \tau_{a_k} \rangle = \sum_{i=1}^{n} \langle \tau_{a_i-1} \prod_{k \neq i} \tau_{a_k} \rangle$$

Proof. The proof is identical to the algebro-geometric one: We have to compute the degree of the intersection product $\prod_{k=1}^{n} \psi_k^{a_k} \cdot \mathcal{M}_{n+1}$. First we replace each term $\psi_k^{a_k}$ ($k \neq 0$) by $\mathrm{ft}_0^*(\psi_k)^{a_k} + \mathrm{ft}_0^*(\psi_k)^{a_k-1} \cdot \varphi_{0,k}$, using lemma 2.1.19 (b), and multiply the product out. As $\varphi_{0,k} \cdot \varphi_{0,k'} = 0$ for $k \neq k'$ (see lemma 2.1.9), we only get the following $n+1$ terms:

$$\prod_{k=1}^{n} \mathrm{ft}_0^*(\psi_k)^{a_k} \cdot \mathcal{M}_{n+1} + \sum_{i=1}^{n} \mathrm{ft}_0^*(\psi_i)^{a_i-1} \cdot \prod_{k \neq i} \mathrm{ft}_0^*(\psi_k)^{a_k} \cdot \varphi_{0,i} \cdot \mathcal{M}_{n+1}$$

Now we push this cycle forward along ft_0 and use the projection formula. The first term vanishes for dimension reasons and, as $\varphi_{0,i}$ pushes forward to $\mathcal{M}_n$ by lemma 2.1.20, the other terms provide the desired result. $\square$

Remark 2.1.24

As in the classical case, the string equation suffices to compute all intersection numbers of Psi-divisors of abstract curves (see [Ko, 1.5.1]). Namely, if $\sum a_i = n - 3$, the equation

$$\langle \tau_{a_1} \cdots \tau_{a_n} \rangle = \frac{(n-3)!}{a_1! \cdots a_n!}$$

holds. This was proven in [KM07, 4.2] using the paper's main theorem [KM07, 4.1] (cited here in theorem 2.1.16). Note, however, that in order to prove the string equation it was not necessary to use [KM07, 4.1].

Lemma 2.1.25
Let $n > 4$ and let $\mathrm{ft}_0 : \mathcal{M}_{n+1} \to \mathcal{M}_n$ be the morphism that forgets the last leaf. Then

$$\mathrm{ft}_{0*}(\mathrm{div}(\varphi_{I|J})) = \begin{cases} \mathcal{M}_n & \text{if } I = \{0, k\} \text{ or } J = \{0, k\} \text{ for some } k \in [n], \\ 0 & \text{otherwise.} \end{cases}$$

Proof. The first part is shown in lemma 2.1.20. So let us prove the second part: First, we choose $i \in I$ and $j \in J$, both different from 0. Consider a facet σ' in $\mathcal{M}_n$ corresponding to a combinatorial type where x_i and x_j are adjacent to the same 3-valent vertex ν. All ridges in $\mathcal{M}_{n+1}$ mapping onto σ' are obtained by attaching x_0 to any of the vertices. If not attached to ν, the induced partition A, B, C, D cannot separate i and j. If attached to ν, the induced partition is $\{0\}, \{i\}, \{j\}, D$. It follows from $\{0, i\} \neq I$ and $\{0, j\} \neq J$ that D intersects both I and J and therefore none of these types is contained in $\mathrm{div}(\varphi_{I|J})$. Hence σ' is not contained in the push forward of $\mathrm{div}(\varphi_{I|J})$. But $\mathcal{M}_n$ is irreducible, thus $\mathrm{ft}_{0*}(\mathrm{div}(\varphi_{I|J})) = 0$. □

Lemma 2.1.26 (cf. [Ko] 1.5.2)
For $n \geq 4$ the equation

$$\mathrm{div}(\psi_1) = \sum_{\substack{I|J \\ 1 \in I;\, 2,3 \in J}} \mathrm{div}(\varphi_{I|J})$$

holds.

Of course, we can replace $1, 2, 3$ by any other choice of distinct leaves.

Proof. We use induction on the number of leaves n. For $n = 4$, only the partition $\{1,4\}|\{2,3\}$ contributes to the sum. But $\mathrm{div}(\psi_1)$ as well as $\mathrm{div}(\varphi_{1,4|2,3})$ is just the single vertex in $\mathcal{M}_4$ parameterizing the curve $\frac{1}{4}\times\frac{2}{3}$ with weight 1. For the induction step, assume $n \geq 4$ and consider the morphism $\mathrm{ft}_0 : \mathcal{M}_{n+1} \to \mathcal{M}_n$ that forgets the leaf x_0 and let $I'|J'$ be a partition of $[n]$. Then $\mathrm{ft}_0^*(\varphi_{I'|J'})$ measures the sum of the lengths of the edges separating I' and J' if present. Hence we obtain

$$\mathrm{ft}_0^*(\varphi_{I'|J'}) = \varphi_{I'\cup\{0\}|J'} + \varphi_{I'|J'\cup\{0\}}.$$

Using the induction hypothesis, we conclude that $\mathrm{ft}_0^*(\psi_1)$ equals the sum on the right hand side except for the partition $\{0,1\}|\{0,1\}^c$. This missing summand is provided by $\mathrm{div}(\psi_1) = \mathrm{div}(\mathrm{ft}_0^*(\psi_1)) + \mathrm{div}(\varphi_{0,1})$ (cf. lemma 2.1.18). □

Lemma 2.1.27 (cf. [Ko] 1.6.1)
Let $n \geq 4$ and let $\mathrm{ft}_0 : \mathcal{M}_{n+1} \to \mathcal{M}_n$ be the morphism that forgets the leaf x_0. Then

$$\mathrm{ft}_{0*}(\mathrm{div}(\psi_0)) = (n-2)\mathcal{M}_n.$$

Proof. We express ψ_0 as a sum of boundary divisors according to lemma 2.1.26 and use the linearity of the push forward. Lemma 2.1.25 says that we get *one* $\mathcal{M}_n$ for each $\varphi_{\{0,k\}|\{0,k\}^c}$ and zero for each other $\varphi_{I|J}$. As k runs through $\{3, \ldots, n\}$, the statement follows. □

Proposition 2.1.28 (Dilaton equation for abstract curves, cf. [Ko] 1.6.2)
Let $\langle \prod_{k=1}^n \tau_{a_k} \rangle$ be a zero-dimensional intersection product. Then

$$\langle \tau_1 \cdot \prod_{k=1}^n \tau_{a_k} \rangle = (n-2)\langle \prod_{k=1}^n \tau_{a_k} \rangle.$$

Proof. The proof is identical to the algebro-geometric one, using lemmata 2.1.19, 2.1.17, 2.1.20, 2.1.27 and the projection formula. We include it for the reader's convenience.

As degree is preserved, we push forward $(\tau_1 \cdot \prod_{k=1}^{n} \tau_{a_k})$ along the forgetful morphism ft_0 forgetting the extra leaf x_0 corresponding to the factor τ_1. To see what happens, we use lemma 2.1.19 (b) and replace each term $\psi_k^{a_k}$ by $\mathrm{ft}_0^*(\psi_k)^{a_k} + \mathrm{ft}_0^*(\psi_k)^{a_k-1} \cdot \varphi_{0,k}$. When we multiply the whole product out, all summands containing a factor $\varphi_{0,k}$ vanish when multiplied with ψ_0 (see lemma 2.1.17). It follows that

$$\psi_0 \cdot \prod_{k=1}^{n} \psi_k^{a_k} = \psi_0 \cdot \prod_{k=1}^{n} \mathrm{ft}_0^*(\psi_k)^{a_k}$$

and the projection formula together with $\mathrm{ft}_{0*}(\mathrm{div}(\psi_0)) = (n-2)\mathcal{M}_n$ from lemma 2.1.27 gives the desired result.

$\square$

2.2 The moduli space of parameterized rational tropical curves

The idea of this section is similar to the previous one. We recall the construction of the moduli space of parameterized tropical rational curves as a tropical variety and prove the tropical analogues of the classically well-known string, dilaton and divisor equations for stable maps. Moreover, we extend the family property of ft_0 to parameterized curves.

2.2.1 Parameterized curves

A *(labelled) degree* Δ in $\mathbb{R}^r$ is a finite set (of labels) Δ together with a map $\Delta \to \mathbb{Z}^r \setminus \{0\}$ to the set of non-zero integer vectors, such that the images of this map, denoted by $v(x_i), i \in \Delta$ as they will later play the role of the directions of the leaves x_i, sum up to zero, i.e. $\sum_{i \in \Delta} v(x_i) = 0$. The number of elements in Δ is denoted by $\#\Delta$ (to distinguish it from the support of a cycle). As an example, we define the *projective degree d (in dimension r)* to be the set $[(r+1)d]$ with the map

$$[(r+1)d] \to \mathbb{Z}^r \setminus \{0\},$$
$$1, \ldots, d \mapsto -e_0,$$
$$d+1, \ldots, 2d \mapsto -e_1,$$
$$\vdots \qquad \vdots$$
$$rd+1, \ldots, (r+1)d \mapsto -e_r,$$

where, as usual, $e_1, \ldots, e_r$ denote the standard basis vectors and $e_0 := -e_1 - \ldots - e_r$.

Definition 2.2.1 (Parameterized curves)

An *n-marked (labelled) parameterized tropical curve of degree Δ in $\mathbb{R}^r$* is a tuple (C, h), where C is an $[n] \dot\cup \Delta$-marked smooth abstract curve and $h : C \to \mathbb{R}^r$ is a tropical morphism such that for all leaves x_i the ray $h(x_i) \subseteq \mathbb{R}^r$ has direction $v(x_i)$, i.e. the primitive generator of x_i is mapped to $v(x_i)$ by the linear part of h on x_i. Here $v(x_i)$ is set to be zero if $i \in [n]$, which implies that marked leaves $x_i, i \in [n]$ are contracted to a point. The genus of (C, h) is defined to be the genus of C.

Remark 2.2.2

The leaves $x_i, i \in [n]$ are called *marked leaves*, as they correspond to the marked points of stable maps classically. Marked leaves are contracted by h. In contrast to that we call the leaves $x_i, i \in \Delta$ *non-contracted leaves*. Our curves are called "labelled" as also the non-contracted leaves are labelled.

Two parameterized curves (C, h) and (C', h') are called isomorphic (and are therefore identified in the following) if there exists an isomorphism $\Phi : C \to C'$ identifying the labels and satisfying $h = h' \circ \Phi$.

Let us compare our definition to [GKM07, definition 4.1]. Conditions (a) and (b) in that definition make sure that h is a tropical morphism in our sense. As condition (c) is also contained in our definition, the two definitions coincide.

Let $\mathcal{M}_n^{\mathrm{lab}}(\mathbb{R}^r, \Delta)$ be the moduli space that parameterizes rational n-marked labelled parameterized tropical curves of degree Δ in $\mathbb{R}^r$. Its construction as a tropical cycle can be found in [GKM07, 4.7], we recall the important facts here: After fixing one of the marked leaves x_i as *anchor leaf* (in [GKM07], this is called "root vertex"; we replace "vertex" by the more precise "leaf" and avoid the combination "root leaf", as it might be a bit confusing from the botanic point of view), we can identify $\mathcal{M}_n^{\mathrm{lab}}(\mathbb{R}^r, \Delta)$ with $\mathcal{M}_{[n]\cup\Delta} \times \mathbb{R}^r$, where the first factor parameterizes the abstract curve C and the second factor contains the coordinates of the image point of the anchor leaf x_i. This suffices to determine the morphism h, as Δ fixes the directions of all leaves of C and therefore, using the balancing condition recursively, all directions of the bounded edges as well (cf. equation 2.1). Hence h is uniquely determined by the lengths of the edges and the coordinates of one image point (in our case $h(x_i)$).

So again, cones in $\mathcal{M}_n^{\mathrm{lab}}(\mathbb{R}^r, \Delta)$ correspond to combinatorial types of the underlying abstract curves, but this time the minimal cone is not zero- but r-dimensional, due to moving $h(x_i)$, and thus the whole curve, in $\mathbb{R}^r$.

For enumerative purposes, we would like to identify curves whose only difference is the labelling of the non-contracted leaves. Let $\mathcal{M}_n(\mathbb{R}^r, \Delta)$ denote the set of these *unlabelled* curves. Then the maximum of elements in a fibre of the map $\mathcal{M}_n^{\mathrm{lab}}(\mathbb{R}^r, \Delta) \to \mathcal{M}_n(\mathbb{R}^r, \Delta)$ forgetting the labelling of the non-contracted leaves is

$$\Delta! := \prod_{v \in \mathbb{Z}^r \setminus \{0\}} n(v)!,$$

where $n(v)$ denotes the number of times v occurs as $v(x_i), i \in \Delta$ (and we assume $n > 1$). This number equals the number of possibilities to label an unlabelled *automorphism-free* curve. Therefore we expect resp. define that each enumerative invariant computed on $\mathcal{M}_n^{\mathrm{lab}}(\mathbb{R}^r, \Delta)$ must be divided by $\Delta!$ to get the corresponding one in $\mathcal{M}_n(\mathbb{R}^r, \Delta)$.

In this section, $I|J$ denotes a (non-empty) partition of $[n] \cup \Delta$ (or $\{0\} \cup [n] \cup \Delta$ if we work with $\mathcal{M}_{n+1}^{\mathrm{lab}}(\mathbb{R}^r, \Delta)$). Again, such partitions can be used as global labels of the edges of our curves. The direction of the image of the corresponding edge under h is given by

$$v_{I|J} := \sum_{i \in I} v(x_i) = -\left(\sum_{j \in J} v(x_j)\right) \tag{2.1}$$

(as an exception, commuting I and J makes a little difference here: $v_{I|J} = -v_{J|I}$). We call $I|J$ *reducible* if $|I|, |J| > 1$ and $v_{I|J} = 0$ (i.e. if the corresponding edge is contracted). This is equivalent to requiring that the corresponding split sets $\Delta_I = I \cap \Delta$ and $\Delta_J = J \cap \Delta$ fulfill the balancing condition, i.e. are degrees on its own. Also the marked leaves split up into $[n] = (I \cap [n]) \cup (J \cap [n])$. That corresponds to the data that is needed to specify an irreducible boundary divisor of the classical moduli space of stable maps $\overline{M}_{0,n}(\mathbf{X}, \beta)$ (for simplicity, we assume that $\mathbf{X}$ is projective and homogenous). Namely, in the this case an irreducible component of the boundary is given by a partition $(A, \beta'|B, \beta'')$ of the marked points $[n] = A \cup B$ and a splitting of the cohomology class $\beta = \beta' + \beta''$ (cf. [KV07, section 2.7.1]). The only difference is that, due to the labelling of the non-contracted leaves, several partitions $I|J$ induce the same degree splitting in the tropical setting.

In contrast to that, the *non-reducible* partitions $I|J$ with $v_{I|J} \neq 0$ do not have a counterpart in the algebro-geometric moduli space.

Remark 2.2.3

Note that the construction $\mathcal{M}_n^{\mathrm{lab}}(\mathbb{R}^r, \Delta) \cong \mathcal{M}_{[n] \cup \Delta} \times \mathbb{R}^r$ has a classical counterpart, as well. Let us explain this in the case of the moduli space $\overline{M}_{0,n}(\mathbb{P}^2, d)$ of n-pointed rational stable maps in $\mathbb{P}^2$ of degree d (following

[KV07, section 2.4]): Let l_0, l_1, l_2 denote the three coordinate lines in $\mathbb{P}^2$ and consider the open set $U \subseteq \overline{M}_{0,n}(\mathbb{P}^2, d)$ of stable maps $\mu : C \to \mathbb{P}^2$ that intersects these three lines transversely (precisely, the inverse image divisor of the divisor $l_0 + l_1 + l_2$ must consist of $3d$ distinct and non-special points $q_{01}, \ldots, q_{0d}, q_{11}, \ldots, q_{1d}, q_{21}, \ldots, q_{2d}$). Every stable map in U induces an $(n + 3d)$-pointed stable curve $\tilde{C}$ by forgetting μ and transforming the q_{ij} into marked points (stability is easy to check). Of course, there is an ambiguity here, as the inverse image points q_{ij} of l_i do not come with a canonical ordering.

The curve $\tilde{C}$ is not an arbitrary curve in $\overline{M}_{0,n+3d}$, but satisfies the property that the three divisors $D_i := q_{i1} + \ldots + q_{id}$ provided by the additional marked points are rationally equivalent as $\mathcal{O}_{\tilde{C}}(D_i) = \mu^* \mathcal{O}_{\mathbb{P}^2}(l_i) = \mu^* \mathcal{O}_{\mathbb{P}^2}(1)$. Let $W \subseteq \overline{M}_{0,n+3d}$ denote the open subset of curves with this property, and let us see how the opposite way works. For any curve $\tilde{C} \in W$, the three divisors D_i arise from sections of the same line bundle $\mathcal{O}_{\tilde{C}}(D_i)$, which induces a morphism $\tilde{\mu} : \tilde{C} \to \mathbb{P}^2$ (the sections do not vanish simultaneously, since the marked points of $\tilde{C}$ are distinct). Hence, by forgetting the additional marked points, we get a stable map $\mu : C \to \mathbb{P}^2$ which intersects the lines transversely in the points q_{ij}. However, the choice of the sections and hence the map μ is not unique. This can also be seen the other way around: The action of the open dense torus T^2 on $\mathbb{P}^2$ also induces an action on U, but the intersection points of a stable map $\mu : C \to \mathbb{P}^2$ with the coordinate lines and therefore the associated $(n + 3d)$-marked curve $\tilde{C}$ are invariant under this action.

This can be summarized as follows. The open set $U \subseteq \overline{M}_{0,n}(\mathbb{P}^2, d)$ is isomorphic to a T^2-bundle over the open set $W \subseteq \overline{M}_{0,n+3d}$ divided by the action of three copies of the symmetric group S_d on $[d]$ (permuting the three sets of marked points $q_{i1}, \ldots, q_{in}$). This observation is helpful classically as, due to the fact that the lines l_0, l_1, l_2 can be replaced by any other choice of three generic lines, the open sets of the form U cover $\overline{M}_{0,n}(\mathbb{P}^2, d)$ (for every stable map $\mu : C \to \mathbb{P}^2$, choose three lines such that they are not tangent to $\mu(C)$ nor meet the image of a special or ramification point). This shows for example that the open subset $\overline{M}_{0,n}^*(\mathbb{P}^2, d) \subseteq \overline{M}_{0,n}(\mathbb{P}^2, d)$ of stable maps without automorphisms is smooth (as on this set the three copies of S_d act freely).

The construction $\mathcal{M}_n^{\text{lab}}(\mathbb{R}^2, d) \cong \mathcal{M}_{[n+3d]} \times \mathbb{R}^2$ via choosing an anchor leaf is the tropical analogue of the above description of the open set W. The tropicalization of the torus T^2 is just $\mathbb{R}^2$, the torus action boils down to translations of the curves, explicitly given by the coordinates of the anchor leaf. The non-contracted leaves (as many as d in any of the directions $-e_0, -e_1, -e_2$) correspond to the intersections of the curve with the three (not existing) boundary lines; and as in the classical setting an ordering of these boundary intersections is necessary when identifying with $\mathcal{M}_{[n+3d]} \times \mathbb{R}^2$.

But then, this motivation of the construction of $\mathcal{M}_n^{\text{lab}}(\mathbb{R}^r, \Delta)$ also reveals difficulties. Indeed, the construction of $\mathcal{M}_n^{\text{lab}}(\mathbb{R}^2, d)$ is related to the compactification $\overline{M}_{0,n} \times \mathbb{P}^2$ of $W \times T^2$ in $\mathbf{X}(\mathcal{M}_{[n+3d]} \times \mathcal{L}_2^2)$, and *not* to the compactification $\overline{M}_{0,n}(\mathbb{P}^2, d)$ of U. The occurrence of non-reducible partitions $I|J$ to which no analogues in $\overline{M}_{0,n}(\mathbb{P}^2, d)$ exist already gives an idea of this problem. However, as up to now no alternative to the given construction of $\mathcal{M}_n^{\text{lab}}(\mathbb{R}^r, \Delta)$ is known, in this thesis we just try to cope with the difficulties arising from this issue. For example, later on we have to make sure that no difference between the tropical and classical WDVV equations arises from the existence of non-reducible partitions (which in existing literature is contained in proving the existence of a contracted edge when the $\mathcal{M}_4$-coordinate is arbitrarily large, see [GM05, 5.1] and [MR08, 4.4]).

Note that there exists a forgetful map $\text{ft}' : \mathcal{M}_n^{\text{lab}}(\mathbb{R}^r, \Delta) \to \mathcal{M}_{[n] \cup \Delta}$ forgetting just the position of the curve in $\mathbb{R}^r$. This forgetful map ft' is a morphism of tropical varieties, as after choosing an anchor leaf and identifying $\mathcal{M}_n^{\text{lab}}(\mathbb{R}^r, \Delta)$ with $\mathcal{M}_{[n] \cup \Delta} \times \mathbb{R}^r$, ft' is just the projection onto the first factor. We use ft′ to define boundary and Psi-divisors on $\mathcal{M}_n^{\mathrm{lab}}(\mathbb{R}^r, \Delta)$.

Definition 2.2.4 (Psi-functions for parameterized curves)
For a partition $I|J$ of $[n] \cup \Delta$ we define the function $\varphi_{I|J}$ on $\mathcal{M}_n^{\mathrm{lab}}(\mathbb{R}^r, \Delta)$ to be $\mathrm{ft}'^*(\varphi_{I|J}^{\mathrm{abstr}})$, where $\varphi_{I|J}^{\mathrm{abstr}}$ is the corresponding function on $\mathcal{M}_{[n] \cup \Delta}$.

For $i = 1, \ldots, n$ we define *the k-th Psi-function on* $\mathcal{M}_n^{\mathrm{lab}}(\mathbb{R}^r, \Delta)$ to be $\psi_k := \mathrm{ft}'^*(\psi_k^{\mathrm{abstr}})$, where the ψ_k^{abstr} is the k-th Psi-function on $\mathcal{M}_{[n] \cup \Delta}$.

Remark 2.2.5
Again, in spite of defining functions, we are actually interested in their divisors. Note that by remark 2.1.10 the pull backs of the respective divisors do not depend on the particular functions. In continuation of remark 2.2.3, let us also mention that it is not obvious from some tropicalization argument that this definition of Psi-classes/Psi-divisors is the "correct" one. Indeed, later on we will have to impose certain restrictions to the use of Psi-divisors (we use them only in connection with a point condition at the marked leaf in question). Another approach by Mark Gross [Gr09], who uses a varying definition (in a completely different framework, however) suggests that further research is necessary here.

We can immediately generalize statement 2.1.16 to the case of parameterized curves.

Lemma 2.2.6 (Products of Psi-divisors for parameterized curves)
Let $a_1, \ldots, a_n$ be positive integers and let $X = \prod_{k=1}^{n} \psi_k^{a_k} \cdot \mathcal{M}_n^{\mathrm{lab}}(\mathbb{R}^r, \Delta)$ be a product of Psi-divisors. Then X is the subfan of $\mathcal{M}_n^{\mathrm{lab}}(\mathbb{R}^r, \Delta)$ consisting of the closure of the cones of dimension $n + \#\Delta - 3 - \sum_{i=1}^{n} a_i$ whose interior curves C have the property:

Let $k_1, \ldots, k_q \in [n]$ be the marked leaves adjacent to a vertex ν of C. Then the valence of ν is

$$\mathrm{val}(\nu) = a_{k_1} + \ldots + a_{k_q} + 3.$$

Let us define the multiplicity of this vertex to be $\mathrm{mult}(\nu) := \binom{\mathrm{val}(\nu)-3}{a_{k_1},\ldots,a_{k_q}}$. *Then the weight of such a cone σ in X is*

$$\omega_X(\sigma) = \prod_\nu \mathrm{mult}(\nu),$$

where the product runs through all vertices ν of an interior curve of σ.

In particular, $\mathrm{div}(\psi_k), k \in [n]$ *consists of all ridges where k is adjacent to the special 4-valent vertex, and all such ridges carry the weight 1.*

Proof. Choose an anchor leaf and identify $\mathcal{M}_n^{\mathrm{lab}}(\mathbb{R}^r, \Delta)$ with $\mathcal{M}_{[n]\cup\Delta} \times \mathbb{R}^r$. Then ft$'$ is just the projection on the first factor and we can apply lemma 1.5.4, i.e. instead of intersecting the pull backs, we can as well intersect on the first factor and then multiply with $\mathbb{R}^r$. Thus,

$$X = (\prod_{k=1}^n (\psi_k^{\mathrm{abstr}})^{a_k} \cdot \mathcal{M}_{[n]\cup\Delta}) \times \mathbb{R}^r,$$

where here ψ_k^{abstr} denotes a Psi-function on $\mathcal{M}_{[n]\cup\Delta}$. Now, as the weight of $\mathbb{R}^r$ is one and the combinatorics of a curve do not change under ft$'$, the statements follows from theorem 2.1.16. $\square$

In the same fashion, the following corollaries of the respective statements for abstract curves can be verified. Again, as the following intersection products are always computed on $\mathcal{M}_n^{\mathrm{lab}}(\mathbb{R}^r, \Delta)$ (resp. $\mathcal{M}_{n+1}^{\mathrm{lab}}(\mathbb{R}^r, \Delta)$), we omit the term "$\cdot\mathcal{M}_n^{\mathrm{lab}}(\mathbb{R}^r, \Delta)$" (resp. "$\cdot\mathcal{M}_{n+1}^{\mathrm{lab}}(\mathbb{R}^r, \Delta)$").

Lemma 2.2.7

Let ft$_0$ be the map $\mathcal{M}_{n+1}^{\mathrm{lab}}(\mathbb{R}^r, \Delta) \to \mathcal{M}_n^{\mathrm{lab}}(\mathbb{R}^r, \Delta)$ that forgets the extra leaf x_0 and assume $n + \#\Delta \geq 4$ (and $n \geq 1$). Furthermore, let x_i, x_j, x_k be pairwise different leaves from $[n]$ and let a be a positive integer. Then the following equations hold:

(a) (cf. [Ko] 2.1.9)

$$\varphi_{i,j} \cdot \varphi_{i,k} = 0$$

(b) (cf. [Ko] 2.1.17)

$$\varphi_{i,j} \cdot \psi_i = 0$$

(c) (cf. [Ko] 2.1.18)

$$\mathrm{div}(\psi_k) = \mathrm{div}(\mathrm{ft}_0^* \psi_k) + \mathrm{div}(\varphi_{0,k})$$

(d) (cf. [Ko] 2.1.19 (a))

$$\varphi_{0,k}^2 = -\mathrm{ft}_0^*(\psi_k) \cdot \varphi_{0,k}$$

(e) (cf. [Ko] 2.1.19 (b))

$$\psi_k^a = \mathrm{ft}_0^*(\psi_k)^a + \mathrm{ft}_0^*(\psi_k)^{a-1} \cdot \varphi_{0,k}$$

(f) (cf. [Ko] 2.1.19 (c))

$$\psi_k^a = \mathrm{ft}_0^*(\psi_k)^a + (-1)^{a-1}\varphi_{0,k}^a$$

(g) (cf. [Ko] 2.1.20)

$$\mathrm{ft}_{0*}(\mathrm{div}(\varphi_{0,k})) = \mathrm{ft}_{0*}(\mathrm{div}(\psi_k)) = \mathcal{M}_n^{\mathrm{lab}}(\mathbb{R}^r, \Delta)$$

(h) (cf. [Ko] 2.1.25)

$$\mathrm{ft}_{0*}(\mathrm{div}(\varphi_{I|J})) = \begin{cases} \mathcal{M}_n^{\mathrm{lab}}(\mathbb{R}^r, \Delta) & \textit{if } I = \{0, k\} \textit{ or } J = \{0, k\} \\ & \textit{for some } k \in [n], \\ 0 & \textit{otherwise.} \end{cases}$$

(i)   (cf. [Ko] 2.1.26)

$$\operatorname{div}(\psi_i) = \sum_{\substack{I|J \\ i \in I; j,k \in J}} \operatorname{div}(\varphi_{I|J}),$$

where the sum runs also through non-reducible *partitions.*

(j)   (cf. [Ko] 2.1.27)

$$\operatorname{ft}_{0*}(\operatorname{div}(\psi_0)) = (n + \#\Delta - 2)\mathcal{M}_n^{\mathrm{lab}}(\mathbb{R}^r, \Delta).$$

Note that this equation is different *to the algebro-geometric analogue, where the factor is* $n - 2$ *(as in the abstract case).*

Proof. As in the proof of lemma 2.2.6, we apply lemma 1.5.4 to the morphism $\operatorname{ft}' : \mathcal{M}_n^{\mathrm{lab}}(\mathbb{R}^r, \Delta) = \mathcal{M}_{[n] \cup \Delta} \times \mathbb{R}^r \to \mathcal{M}_{[n] \cup \Delta}$ forgetting the position in $\mathbb{R}^r$. This means that instead of computing the intersection products on $\mathcal{M}_n^{\mathrm{lab}}(\mathbb{R}^r, \Delta)$ we can compute them on $\mathcal{M}_{[n] \cup \Delta}$ and therefore use the corresponding statements for abstract curves. For statements (c) – (h) and (j) we also use $\operatorname{ft}_0 = \operatorname{ft}_0^{\mathrm{abstr}} \times \operatorname{id}$.   □

Definition 2.2.8 (Evaluation maps and their pull backs)
The *evaluation map* $\operatorname{ev}_k : \mathcal{M}_n^{\mathrm{lab}}(\mathbb{R}^r, \Delta) \to \mathbb{R}^r$, for $k \in [n]$, maps each parameterized curve (C, h) to the position of its k-th leaf $h(x_k)$ (see [GKM07, 4.2]). If we choose one of the marked leaves, say x_a, as anchor leaf, then the evaluation maps are morphisms from $\mathcal{M}_{[n] \cup \Delta} \times \mathbb{R}^r$ to $\mathbb{R}^r$ obeying the following mapping rule:

$$(C^{\mathrm{abstr}}, P) \mapsto P + \sum_{\substack{I|J \\ a \in I, k \in J}} \varphi_{I|J}(C^{\mathrm{abstr}}) v_{I|J}$$

In particular, if our anchor leaf is chosen to be x_k, then ev_k is just the projection onto the second factor. Let $C \in Z_d(\mathbb{R}^r)$ be a complete intersection given by $C = \varphi_1 \cdots \varphi_l \cdot \mathbb{R}^r$. Then by remark 2.1.10 there exists a well-defined *pull back of C along* ev_k

$$\mathrm{ev}_k^*(C) := \mathrm{ev}_k^*(\varphi_1) \cdots \mathrm{ev}_k^*(\varphi_l).$$

This can be extended to sums of complete intersections.

Proposition 2.2.9 (Family property of $\mathrm{ft}_0, \mathrm{ev}_0$ for parameterized curves)
Let p be a point in $\mathcal{M}_n^{\mathrm{lab}}(\mathbb{R}^r, \Delta)$ and let $C_p = \mathrm{ft}_0^{-1}(p)$ be the fibre of p under the forgetful morphism $\mathrm{ft}_0 : \mathcal{M}_{n+1}^{\mathrm{lab}}(\mathbb{R}^r, \Delta) \to \mathcal{M}_n^{\mathrm{lab}}(\mathbb{R}^r, \Delta)$. Then the following holds:

(a) *When we equip all its facets with weight 1, C_p is a rational smooth abstract curve. Its leaves are the naturally $([n] \dot\cup \Delta)$-marked sets $L_i := \overline{\{y \in C_p | \varphi_{0,i}(y) > 0\}}$.*

(b) *The tuple $(C_p, \mathrm{ev}_0 |_{C_p})$ is an n-marked parameterized curve of degree Δ. Moreover, p represents $(C_p, \mathrm{ev}_0 |_{C_p})$.*

(c) *Let $\sum_k \mu_k p_k = \varphi_1 \cdots \varphi_{n+\#\Delta-3} \cdot \mathcal{M}_n^{\mathrm{lab}}(\mathbb{R}^r, \Delta)$ be a zero-dimensional cycle in $\mathcal{M}_n^{\mathrm{lab}}(\mathbb{R}^r, \Delta)$ obtained as the intersection product of convex functions φ_j. Then*

$$\mathrm{ft}_0^*(\varphi_1) \cdots \mathrm{ft}_0^*(\varphi_{n+\#\Delta-3}) \cdot \mathcal{M}_{n+1}^{\mathrm{lab}}(\mathbb{R}^r, \Delta) = \sum_k \mu_k C_{p_k}.$$

We write this as $\mathrm{ft}_0^*(\sum_k \mu_k p_k) = \sum_k \mu_k C_{p_k}$.

Proof. (a): First of all, let us fix an anchor leaf $x_a, a \in [n]$ in order to identify $\mathcal{M}_{n+1}^{\mathrm{lab}}(\mathbb{R}^r, \Delta) = \mathcal{M}_{n+\#\Delta+1} \times \mathbb{R}^r$ and $\mathcal{M}_n^{\mathrm{lab}}(\mathbb{R}^r, \Delta) = \mathcal{M}_{[n] \cup \Delta} \times \mathbb{R}^r$. Again we use $\mathrm{ft}_0 = \mathrm{ft}_0^{\mathrm{abstr}} \times \mathrm{id}$, where $\mathrm{ft}_0^{\mathrm{abstr}}$ is the corresponding forgetful map on the "abstract" space. Then the fibre of $p = (p', P) \in \mathcal{M}_{[n] \cup \Delta} \times \mathbb{R}^r$ equals $C_{p'} \times \{P\}$, where $C_{p'}$ is the $[n] \dot\cup \Delta$-marked rational smooth abstract curve considered in proposition 2.1.21 (a)–(c).

(b): We have to check that the leaves L_i are mapped to rays with "correct" direction $v(x_i)$. For curves corresponding to points in L_i, the only length that varies is that of the third edge adjacent to the same 3-valent vertex as x_i and x_0. Hence we can use the description of ev_0 in 2.2.8 and obtain for all $y \in L_i$

$$\mathrm{ev}_0|_{L_i}(y) = Q + \varphi_{0,i}(y) \cdot v_{\{0,i\}|\{0,i\}^c},$$

where $Q \in \mathbb{R}^r$ is some constant vector. But $v_{\{0,i\}|\{0,i\}^c} = v(x_i) + v(x_0) = v(x_i)$ is the expected direction.

To show that $p = (p', P)$ represents $(C_p, \mathrm{ev}_0|_{|C_p|})$ it actually suffices to prove that the anchor leaf L_a of C_p is mapped to the point P under ev_0, which is obviously the case as $\mathrm{ev}_0|_{L_a} = \mathrm{ev}_a|_{L_a}$ and ev_a is just the projection on the second factor of $C_{p'} \times \{P\}$.

(c): With the help of lemma 2.2.7 (g), we can use literally the same proof as in the abstract case (cf. proposition 2.1.21 (d)). $\qquad\square$

2.2.2 Tropical gravitational descendants and their enumerative meaning

Let us fix the following notation.

Notation 2.2.10 (Tropical gravitational descendants)
We now extend our τ-notation to the case of parameterized curves. For a given labelled degree Δ, integers $a_1, \ldots, a_n$ and (sums of) complete intersections $C_1, \ldots, C_n \in Z_*(\mathbb{R}^r)$ we define

$$(\tau_{a_1}(C_1) \cdots \tau_{a_n}(C_n))_\Delta^{\mathbb{R}^r} := \psi_1^{a_1} \cdot \mathrm{ev}_1^*(C_1) \cdots \psi_n^{a_n} \cdot \mathrm{ev}_n^*(C_n) \cdot \mathcal{M}_n^{\mathrm{lab}}(\mathbb{R}^r, \Delta)$$

(if one the a_k is negative, we define the cycle to be $\emptyset$). Once again, each factor $\tau_{a_k}(C_k)$ stands for a marked leaf x_k *restricted by a_k Psi-conditions and the incidence condition C_k*. In the special case $C_k = \mathbb{R}^r$, no pull back along ev_k occurs in the product. We call x_k *unrestricted* if $a_k = 0$ and $C_k = \mathbb{R}^r$.

Let c_k be the codimension of C_k in $\mathbb{R}^r$. If $\sum(a_k+c_k) = \dim(\mathcal{M}_n^{\mathrm{lab}}(\mathbb{R}^r, \Delta))$ $= n + \#\Delta + r - 3$, the above cycle is zero-dimensional and we denote its degree by

$$\langle \tau_{a_1}(C_1) \cdots \tau_{a_n}(C_n) \rangle_\Delta^{\mathbb{R}^r}.$$

These numbers are called *tropical gravitational descendants* or *tropical descendant Gromov-Witten invariants*.

Before we try to compute these numbers, let us first make explicit what they count in the "generic case". For this we need some tools concerning general position of the incidence conditions and the following remark.

Remark 2.2.11 (Pulling back preserves numerical equivalence)
Let C be a complete intersection in $\mathbb{R}^r$ and let $f : Y \to \mathbb{R}^r$ be a tropical morphism. Then, if C is rationally equivalent to zero/numerically equivalent to zero/ has degree zero in $\mathbb{R}^r$ (cf. section 1.4), also $f^*(C)$ is numerically equivalent to zero. Indeed, set $l = \dim(f^*(C))$ and let $\varphi_1, \ldots, \varphi_l$ be Cartier divisors on Y, then

$$\deg(\varphi_1 \cdots \varphi_l \cdot f^*(C)) = \deg(C \cdot f_*(\varphi_1 \cdots \varphi_l \cdot Y)) = 0$$

holds. Here we used the projection formula and the fact that the push forward of a zero-dimensional cycle preserves degree.

In particular, in order to compute

$$\langle \tau_{a_1}(C_1) \cdots \tau_{a_n}(C_n) \rangle_\Delta^{\mathbb{R}^r},$$

we can replace all C_k by rationally equivalent cycles (translations, for example) without changing the invariant.

We now investigate what we can say about the set-theoretic preimage of a general translation of a cycle under a morphism f.

Lemma 2.2.12

Let $\mathcal{X}$ be a pure-dimensional polyhedral complex and let $f : \mathcal{X} \to \mathbb{R}^r$ be a map which is affine on every cell of $\mathcal{X}$. Furthermore, let $\mathcal{C}$ be a polyhedral complex in $\mathbb{R}^r$ and consider the subcomplex $f^{-1}(\mathcal{C})$ of $\mathcal{X}$ consisting of all polyhedra $\tau \cap f^{-1}(\gamma), \tau \in \mathcal{X}, \gamma \in \mathcal{C}$. Then for a general translation $\mathcal{C}' = \mathcal{C} + v$ (i.e. $v \in \mathbb{R}^r$ can be chosen from an open dense subset of $\mathbb{R}^r$) the codimension of each non-empty polyhedron $\tau \cap f^{-1}(\gamma)$ of $\mathcal{X}$ is equal to

$$\mathrm{codim}_{\mathcal{X}}(\tau \cap f^{-1}(\gamma)) = \mathrm{codim}_{\mathcal{X}}(\tau) + \mathrm{codim}_{\mathbb{R}^r}(\gamma).$$

Proof. For each τ in $\mathcal{X}$ and γ in $\mathcal{C}$ we consider the affine map

$$f_\tau : \mathrm{AffiSpan}(\tau) \to \mathbb{R}^r,$$

induced by $f|_\tau$ (where AffiSpan enotes the affine span of τ). Now we are interested in $\tau \cap f^{-1}(\gamma') = \tau \cap f_\tau^{-1}(\gamma')$ for general translations γ' of γ. We have to distinguish the cases $\mathrm{Im}(f_\tau) + V_\gamma = \mathbb{R}^r$ and $\mathrm{Im}(f_\tau) + V_\gamma \neq \mathbb{R}^r$. In the latter case, $f_\tau^{-1}(\gamma')$ is empty for general γ'. In the former case, $f_\tau^{-1}(\gamma')$ is a polyhedron of dimension $\dim(\tau) + \dim(\gamma) - r$, and for general γ' it is disjoint from τ or intersects the interior of τ, in which case $\tau \cap f_\tau^{-1}(\gamma')$ has the dimension $\dim(\tau) - \mathrm{codim}_{\mathbb{R}^r}(\gamma)$, which is the expected dimension.

As there are only finitely many pairs τ, γ this holds simultaneously for all pairs for general enough translations of $\mathcal{C}$. $\square$

This technical statement has the following more useful consequences:

Corollary 2.2.13 (Preimages of general translations)

Let $\mathcal{X}$ be a polyhedral complex and let $f_k : \mathcal{X} \to \mathbb{R}^r, k = 1, \ldots, n$ be maps which are affine on the cells of $\mathcal{X}$. Moreover, let $\mathcal{C}_k, k = 1, \ldots, n$ be polyhedral complexes in $\mathbb{R}^r$. Then for general translations $\mathcal{C}'_k = \mathcal{C}_k + v_k, v_k \in \mathbb{R}^r$ the following holds: Either $\mathcal{Z} := f_1^{-1}(\mathcal{C}'_1) \cap \ldots \cap f_n^{-1}(\mathcal{C}'_n)$ is empty or

(a) the codimension of $\mathcal{Z}$ in $\mathcal{X}$ equals the sum

$$\operatorname{codim}_{\mathcal{X}}(\mathcal{Z}) = \sum_{k=1}^{n} \operatorname{codim}_{\mathbb{R}^r}(\mathcal{C}_k),$$

(b) $\mathcal{Z}$ is pure-dimensional,

(c) if a cell α of $\mathcal{Z}$ is contained in a cell τ of $\mathcal{X}$, the codimensions satisfy $\operatorname{codim}_{\mathcal{X}}(\tau) \leq \operatorname{codim}_{\mathcal{Z}}(\alpha)$ (in particular, the interior of a facet of $\mathcal{Z}$ is contained in the interior of a facet of $\mathcal{X}$),

(d) if the images $f_k(\alpha)$ of a polyhedron α of $\mathcal{Z}$ are contained in polyhedra γ_k of $\mathcal{C}_k$, the codimensions satisfy $\sum_{k=1}^{n} \operatorname{codim}_{\mathcal{C}_k}(\gamma_k) \leq \operatorname{codim}_{\mathcal{Z}}(\alpha)$.

Proof. It is easy to prove the statement in the case $n = 1$: (a), (b) and (c) are immediate consequences of lemma 2.2.12 and (d) follows from applying lemma 2.2.12 to the $(r - \operatorname{codim}_{\mathcal{Z}}(\alpha) - 1)$-dimensional skeleton of $\mathcal{C}_1$ (if γ_1 belonged to this skeleton, α would be contained in its preimage, which (for general translations) contradicts (a)). Now the statement follows if we apply the case of a single morphism to $f_1 \times \ldots \times f_n : \mathcal{X} \to (\mathbb{R}^r)^n$ and $\mathcal{C} := \mathcal{C}_1 \times \ldots \times \mathcal{C}_n$. $\qquad \square$

Of course, we want to apply the previous statement to products of Psi- and incidence conditions.

Remark 2.2.14 (Enumerative relevance of gravitational descendants) Let $Z = (\tau_{a_1}(C_1) \cdots \tau_{a_n}(C_n))$ be an intersection product as defined above and set $X = \prod_{k=1}^{n} \psi_k^{a_k} \cdot \mathcal{M}_n^{\text{lab}}(\mathbb{R}^r, \Delta)$. We always have the inclusion

$$|Z| \subseteq \operatorname{ev}_1^{-1}(|C_1|) \cap \ldots \cap \operatorname{ev}_n^{-1}(|C_n|) \cap |X|,$$

as $|\operatorname{ev}_k^*(C_1)| = |\mathcal{M}_{n+\#\Delta} \times C_1| = \operatorname{ev}_k^{-1}(|C_1|)$ holds (with respect to anchor leaf x_k). Now we apply corollary 2.2.13 to the morphisms $\operatorname{ev}_k : X \to \mathbb{R}^r$ and conclude that after replacing all the cycles C_k by general translations (called *general conditions* in the following), we can assume that both sets are of the same dimension. Hence Z is basically the set of curves C such that

- every vertex $\nu \in C$ with adjacent marked leaves $k_1, \ldots, k_q$ fulfills $\mathrm{val}(\nu) \geq a_{k_1} + \ldots + a_{k_q} + 3$,

- $\mathrm{ev}_k(C) \in C_k$ holds for all $k = 1, \ldots, n$,

with weights as additional structure.

The most interesting case is when Z is zero-dimensional. In this case, Z is the finite set of curves C such that

- every vertex $\nu \in C$ with adjacent marked leaves $k_1, \ldots, k_q$ fulfills $\mathrm{val}(\nu) = a_{k_1} + \ldots + a_{k_q} + 3$ (here equality is due to corollary 2.2.13 (c)),

- $\mathrm{ev}_k(C) \in C_k$ holds for all $k = 1, \ldots, n$,

and each such curve carries a certain weight $\omega_Z(C)$. This weight can be computed locally on $\mathrm{Star}_X(C) = \omega_X(\sigma) \cdot V_\sigma$, where σ is the facet of X containing C. Hence $\omega_Z(C)$ is the product of $\omega_X(\sigma)$ and the weight coming from the intersection of the evaluation pull backs on the vector space V_σ. Moreover, it follows from part (d) of corollary 2.2.13 that locally around C all evaluation pull backs can be assumed to be of the form $\mathrm{ev}_k^*(a \cdot \max\{\alpha, 0\})$, where a is an integer and α is an affine form on $\mathbb{R}^r$ (as the neighbourhood of a point in the interior of a facet in C_k is described by (a product of) functions of the form $a \cdot \max\{\alpha, 0\}$). Hence we can use lemma 1.2.9 to compute this second factor as a determinant resp. lattice index (cf. remark 2.4.21).

2.2.3 The string, dilaton and divisor equations

Together with the following remark, the extension of the string and divisor equations to the case of parameterized curves is no problem. Afterwards, we prove a general divisor equation.

Remark 2.2.15
Let $\mathrm{ft}_0 : \mathcal{M}_{n+1}^{\mathrm{lab}}(\mathbb{R}^r, \Delta) \to \mathcal{M}_n^{\mathrm{lab}}(\mathbb{R}^r, \Delta)$ be the morphism that forgets the leaf x_0. Then by abuse of notation the equation

$$\mathrm{ft}_0^*(\mathrm{ev}_k) = \mathrm{ev}_k$$

holds for all $k \in [n]$, i.e. the position of the image of x_k does not change when forgetting x_0.

Theorem 2.2.16 (String equation for parameterized curves, cf. [Ko] 4.3.1)
Let $(\tau_0(\mathbb{R}^r) \cdot \prod_{k=1}^n \tau_{a_k}(C_k))_\Delta$ be a zero-dimensional cycle. Then the equation

$$\langle \tau_0(\mathbb{R}^r) \cdot \prod_{k=1}^n \tau_{a_k}(C_k) \rangle_\Delta = \sum_{k=1}^n \langle \tau_{a_k-1}(C_k) \cdot \prod_{l \neq k} \tau_{a_l}(C_l) \rangle_\Delta.$$

holds.

Theorem 2.2.17 (Dilaton equation for parameterized curves, cf. [Ko] 4.3.1)
Let $(\tau_1(\mathbb{R}^r) \cdot \prod_{k=1}^n \tau_{a_k}(C_k))_\Delta$ be a zero-dimensional cycle. Then the equation

$$\langle \tau_1(\mathbb{R}^r) \cdot \prod_{k=1}^n \tau_{a_k}(C_k) \rangle_\Delta = (n + \#\Delta - 2)\langle \prod_{k=1}^n \tau_{a_k}(C_k) \rangle_\Delta.$$

holds.

Proofs. In both cases, the proofs are completely analogous to the abstract case using lemma 2.2.7 and remark 2.2.15. $\qquad\square$

Remark 2.2.18
Note again that the factor $(n + \#\Delta - 2)$ occurring in the dilaton equation is different from the algebro-geometric factor $n - 2$, due to $\mathrm{ft}_{0*}(\psi_0) =$

$(n + \#\Delta - 2) \cdot \mathcal{M}_n^{\text{lab}}(\mathbb{R}^r, \Delta)$ (cf. 2.2.7 (j)) — this gives an example of the problems mentioned in remark 2.2.3.

Lemma 2.2.19 (cf. [Ko] 5.1.6)
Let φ be a Cartier divisor on $\mathbb{R}^r$. Then

$$\text{ev}_k^*(\varphi) \cdot \varphi_{k,l} \cdot \mathcal{M}_n^{\text{lab}}(\mathbb{R}^r, \Delta) = \text{ev}_l^*(\varphi) \cdot \varphi_{k,l} \cdot \mathcal{M}_n^{\text{lab}}(\mathbb{R}^r, \Delta)$$

Proof. In all curves corresponding to points in $\text{div}(\varphi_{k,l})$, the leaves k and l lie at a common vertex. Therefore their coordinates in $\mathbb{R}^r$ must agree, which means $\text{ev}_k|_{|\text{div}(\varphi_{k,l})|} = \text{ev}_l|_{|\text{div}(\varphi_{k,l})|}$. The result follows. $\qquad\square$

For a given labelled degree Δ, we define $\delta(\Delta)$ to be the associated unlabelled degree in the sense of section 1.4: $\delta(\Delta)$ is the one-dimensional balanced fan in $\mathbb{R}^r$ consisting of all the rays generated by the direction vectors $v_k, k \in \Delta$. The weight of such a ray $\mathbb{R}_{\geq} v$, where v is primitive, is given by

$$\sum_{\substack{k \in \Delta \\ v_k \in \mathbb{Z}_{>0} v}} |\mathbb{Z}v/\mathbb{Z}v_k|.$$

Obviously, if $(C, h) \in \mathcal{M}_n^{\text{lab}}(\mathbb{R}^r, \Delta)$ is an arbitrary n-marked parameterized curve of degree Δ, then by definition $\delta(h(C)) = \delta(\Delta)$ holds.

For a given Cartier divisor φ on $\mathbb{R}^r$ we define $\varphi \cdot \Delta$ to be $\deg(\varphi \cdot \delta(\Delta))$.

Proposition 2.2.20 (cf. [Ko] 5.1.5)
Let φ be a Cartier divisor on $\mathbb{R}^r$ and define $Y := \text{ev}_0^*(\varphi) \cdot \mathcal{M}_{n+1}^{\text{lab}}(\mathbb{R}^r, \Delta)$. Then

$$\text{ft}_{0*}(Y) = (\varphi \cdot \Delta) \cdot \mathcal{M}_n^{\text{lab}}(\mathbb{R}^r, \Delta).$$

Proof. As our moduli space $\mathcal{M}_n^{\text{lab}}(\mathbb{R}^r, \Delta)$ is irreducible, we know that $\text{ft}_{0*}(Y) = \alpha \cdot \mathcal{M}_n^{\text{lab}}(\mathbb{R}^r, \Delta)$ for an integer α. To compute this number, we set $m := n + \#\Delta + r - 3$ and consider the zero-dimensional intersection product $Z = \varphi_1 \cdots \varphi_m \cdot \mathcal{M}_n^{\text{lab}}(\mathbb{R}^r, \Delta)$ of arbitrary convex functions

$\varphi_1, \ldots, \varphi_m$ such that $\deg(Z) \neq 0$ (e.g. $Z = \psi_1^{m-r} \cdot \mathrm{ev}_1(P)$ for some point $P \in \mathbb{R}^r$). If we pull back Z along ft_0, we know by the projection formula that

$$\deg(\mathrm{ev}_0^*(\varphi) \cdot \mathrm{ft}_0^*(Z)) = \alpha \cdot \deg(Z).$$

But then, by the family property of ft_0 we know that Z is the union of the curves represented by the points in Z (with according weights) and therefore the push forward $\mathrm{ev}_{0*}(\mathrm{ft}_0^*(Z))$ is rationally equivalent to its degree

$$\delta(\mathrm{ev}_{0*}(\mathrm{ft}_0^*(Z))) = \deg(Z) \cdot \delta(\Delta).$$

So, applying the projection formula to ev_0, we obtain

$$\deg(\mathrm{ev}_0^*(\varphi) \cdot \mathrm{ft}_0^*(Z)) = \deg(Z) \cdot (\varphi \cdot \Delta).$$

But this implies $\varphi \cdot \Delta = \alpha$, which proves the claim. $\qquad\square$

Theorem 2.2.21 (Divisor equation, cf. [Ko] 4.3.2)
Let φ be a Cartier divisor on $\mathbb{R}^r$ and let $(\prod_{k=1}^n \tau_{a_k}(C_k))_\Delta$ be a one-dimensional cycle. Then the equation

$$\langle \tau_0(\varphi) \cdot \prod_{k=1}^n \tau_{a_k}(C_k) \rangle_\Delta = (\varphi \cdot \Delta) \langle \prod_{k=1}^n \tau_{a_k}(C_k) \rangle_\Delta + \sum_{k=1}^n \langle \tau_{a_k-1}(\varphi \cdot C_k) \prod_{l \neq k} \tau_{a_l}(C_l) \rangle_\Delta.$$

holds.

Proof. First we use lemma 2.2.7 (e) and (a): We replace each factor $\psi_k^{a_k}$ by $\mathrm{ft}_0^*(\psi_k)^{a_k} + \mathrm{ft}_0^*(\psi_k)^{a_k-1} \cdot \varphi_{0,k}$ and multiply out. All terms containing two φ-factors vanish. In terms with only one factor $\varphi_{0,k}$, we replace $\mathrm{ev}_0(\varphi)$ by $\mathrm{ev}_k(\varphi)$ using lemma 2.2.19. Now we push forward along ft_0 and produce the desired equation by applying the projection formula as well as $\mathrm{ft}_{0*}(\mathrm{div}(\mathrm{ev}_0(\varphi))) = (\varphi \cdot \Delta) \cdot \mathcal{M}_n^{\mathrm{lab}}(\mathbb{R}^r, \Delta)$ and $\mathrm{ft}_{0*}(\mathrm{div}(\varphi_{0,k})) = \mathcal{M}_n^{\mathrm{lab}}(\mathbb{R}^r, \Delta)$. $\qquad\square$

2.3 The splitting lemma

The basic fact used to compute intersection invariants of $\overline{M}_{g,n}(X,\beta)$ is the recursive structure of its boundary: Its irreducible components correspond to reducible curves with two components according to a certain partition of the combinatoric data and therefore are (nearly) a product of two "smaller" moduli spaces. In this section we investigate how far this principle can be carried over to the tropical world.

2.3.1 The case of abstract curves

Let S be a finite set. The moduli space of $|S|$-marked tropical curves where we label the leaves by elements in S is denoted by $\mathcal{M}_S$. For each partition $I|J$ of $[n]$ we construct the map $\rho_{I|J} : \mathcal{M}_{I\cup\{x\}} \times \mathcal{M}_{J\cup\{y\}} \to \varphi_{I|J} \cdot \mathcal{M}_n$ by the following rule: Given two curves $(p_I, p_J) \in \mathcal{M}_{I\cup\{x\}} \times \mathcal{M}_{J\cup\{y\}}$, we remove the extra leaves x and y and glue the curves together at the two vertices to which these leaves have been adjacent. We could also say, we glue x and y together by creating a bounded edge whose length we define to be 0. In the coordinates of the space of metrics, this map is given by the linear map

$$\rho_{I|J} : \mathbb{R}^{\binom{I}{2}} \times \mathbb{R}^{\binom{J}{2}} \to \mathbb{R}^{\binom{n}{2}},$$

$$(p_I, p_J) \mapsto p,$$

where

$$p_{k,l} := \begin{cases} (p_I)_{k,l} & \text{if } k,l \in I, \\ (p_J)_{k,l} & \text{if } k,l \in J, \\ (p_I)_{k,x} + p_{J_y,l} & \text{if } k \in I, l \in J. \end{cases}$$

Here be have to be careful: This map does *not* induce a linear map on the corresponding quotients in which our moduli spaces really live and therefore $\rho_{I|J}$ is *not* a tropical morphism of our moduli spaces. This follows from the fact that the image under $\rho_{I|J}$ of a tuple of star metrics is in general *not* a star metric again (only if the lengths of x and y sum up to zero). At least $\rho_{I|J}$ is piecewise linear (i.e. it is linear on all cones of $\mathcal{M}_{I \cup \{x\}} \times \mathcal{M}_{J \cup \{y\}}$). Its image is a polyhedral complex, namely the positive part of $\varphi_{I|J} \cdot \mathcal{M}_n$ (i.e. it consists of all (faces of) facets $^A_D \times ^B_C$ with $A \cup B = I$). So let us make the appropriate definition for this kind of maps.

Definition 2.3.1 (Morphisms of polyhedral complexes)
Let $\mathcal{X}$ and $\mathcal{Y}$ be (rational) polyhedral complexes. Then a *morphism of polyhedral complexes* is a map $\rho : |\mathcal{X}| \to |\mathcal{Y}|$ that satisfies for each polyhedron $\sigma \in \mathcal{X}$

(a) $\rho(\sigma) \in \mathcal{Y}$,

(b) $\rho|_\sigma$ is affine linear,

(c) $\rho(\Lambda_\sigma) \subseteq \Lambda_{\rho(\sigma)}$.

We call ρ an *isomorphism of polyhedral complexes* if there exists an inverse morphism. It other words, an isomorphism is a bijection ρ between $|\mathcal{X}|$ and $|\mathcal{Y}|$ (as well as between $\mathcal{X}$ and $\mathcal{Y}$) and $\rho(\Lambda_\sigma) = \Lambda_{\rho(\sigma)}$ for all $\sigma \in \mathcal{X}$. If $\mathcal{X}$ and $\mathcal{Y}$ are weighted, we add the further requirement that identified weights must have agreeing weights.

Lemma 2.3.2 (Products of Psi-divisors with the boundary)
The facets of the fan $\varphi_{I|J} \cdot \psi_1^{a_1} \cdots \psi_n^{a_n} \cdot \mathcal{M}_n$ with positive weight are precisely the cones σ in $\mathcal{M}_n$ with the following properties:

Consider a curve in the interior of σ. Let $E(\nu) \in [n]$ be the set of leaves adjacent to a vertex ν and let $P(\nu)$ be the $\mathrm{val}(\nu)$-fold partition of $[n]$ obtained by removing ν. Then the following holds:

(a) *There exists one special vertex ν_{spec} whose partition $P(\nu_{spec})$ is a subpartition of $I|J$ and whose valence is $\left(\sum_{k \in E(\nu)} a_k\right) + 4$.*

(b) Let m_I be the number of sets in $P(\nu_{spec})$ contained in I. Then $m_I + 1 = (\sum_{k \in E(\nu) \cap I} a_k) + 3$ (together with (a), the analogue $m_J + 1 = (\sum_{k \in E(\nu) \cap J} a_k) + 3$ follows). In particular, $m_I, m_J > 1$.

(c) The valence of all other vertices ν equals $(\sum_{k \in E(\nu)} a_k) + 3$.

Furthermore, the facets of $\varphi_{I|J} \cdot \psi_1^{a_1} \cdots \psi_n^{a_n} \cdot M_n$ with negative weight fulfill the same properties (a) and (c) and the property

(b') Let m_I (resp. m_J) be the number of sets in $P(\nu_{spec})$ contained in I (resp. J). Then $m_I = 1$ or $m_J = 1$, i.e. $I \in P(\nu_{spec})$ or $J \in P(\nu_{spec})$.

Let us give a short explanation of conditions (b) and (b') by comparing the statement to the case of $\operatorname{div}(\varphi_{I|J})$ (without Psi-divisors, cf. lemma 2.1.7). Then (b') corresponds to a ridge $\frac{A}{D} \times \frac{B}{C}$ with $I = A$, i.e. $m_I = 1$, and this ridge has a negative weight in $\operatorname{div}(\varphi_{I|J})$. Instead, (b) corresponds to a ridge $\frac{A}{D} \times \frac{B}{C}$ with $I = A \cup B$, i.e. $m_I = 2$, and such a ridge has a positive weight in $\operatorname{div}(\varphi_{I|J})$.

Proof. We know how $X := \psi_1^{a_1} \cdots \psi_n^{a_n} \cdot M_n$ looks like by theorem 2.1.16. In the combinatorial type of a facet of X the valence of each vertex is $(\sum_{k \in E(\nu)} a_k) + 3$; in the combinatorial type of a ridge, there is one special vertex ν_{spec} with valence $(\sum_{k \in E(\nu)} a_k) + 4$. The balancing condition of a ridge is given by the equation

$$\sum_{I'|J'} \omega_{I'|J'} V_{I'|J'} = \sum_{\substack{I'|J' \\ I' \in P(V_{spec})}} \lambda_{I'|J'} V_{I'|J'}, \qquad (2.2)$$

where the left hand sum runs through all superpartitions $I'|J'$ of $P(\nu_{spec})$ not appearing in the right hand sum, $\omega_{I'|J'}$ denotes the weight of the facet obtained by inserting an edge $I'|J'$ and $\lambda_{I'|J'}$ is some (rational) coefficient. Therefore the weight ω that this ridge obtains when intersecting X with

$\varphi_{I|J}$ is given by

$$
\omega = \begin{cases} 0 & \text{if } I|J \text{ is } not \text{ a superpartition of } P(V_{\text{spec}}), \\ \lambda_{I|J} & \text{if } I \in P(V_{\text{spec}}) \text{ or } J \in P(V_{\text{spec}}), \\ \omega_{I|J} & \text{otherwise.} \end{cases}
$$

This already shows two implications: As all weights $\omega_{I'|J'}$ are at least non-negative, a ridge can only obtain a negative weight if it fulfills conditions (a), (b') and (c). On the other hand, if a ridge of X satisfies properties (a), (b) and (c), then $\omega_{I|J}$ is positive hence the ridge obtains a positive weight. It remains to show the converse, which can be done by proving that all $\lambda_{I'|J'}$ are non-negative. To see this, we consider equation 2.2 in $\mathbb{R}^{\binom{r}{2}}$ and compare some coordinate entries.

Let K be an arbitrary element of $P(\nu_{\text{spec}})$; we want to show that $\lambda_K := \lambda_{K|K^c}$ is non-negative. We choose two more arbitrary sets L_1, L_2 from the partition $P(\nu_{\text{spec}})$ and fix arbitrary leaves $k \in K$, $l_i \in L_i$. Now the k, l_i-entry of the right hand side of equation 2.2 equals $\lambda_K + \lambda_{L_i}$ and analogously the l_1, l_2-entry equals $\lambda_{L_1} + \lambda_{L_2}$. Therefore, by adding the two k, l_i-entries and subtracting the l_1, l_2-entry we get $2\lambda_K$. Meanwhile, on the left hand side we get

$$
2\lambda_K = \sum_{\substack{I'|J' \\ k \in I' \\ l_1 \in J'}} \omega_{I'|J'} + \sum_{\substack{I'|J' \\ k \in I' \\ l_2 \in J'}} \omega_{I'|J'} - \sum_{\substack{I'|J' \\ l_1 \in I' \\ l_2 \in J'}} \omega_{I'|J'}
$$

$$
= \sum_{I'|J'} \alpha_{I'|J'} \omega_{I'|J'},
$$

where

$$
\alpha_{I'|J'} = \begin{cases} 2 & \text{if } k \in I', \, l_1, l_2 \in J' \\ 0 & \text{if } k, l_1 \in I', \, l_2 \in J' \\ 0 & \text{if } k, l_2 \in I', \, l_1 \in J' \\ 0 & \text{if } k, l_1, l_2 \in I'. \end{cases}
$$

But as all the weights $\omega_{I'|J'}$ are non-negative, it follows that λ_K is non-negative. $\qquad\square$

Let $\mathcal{X}$ be a weighted polyhedral complex, then $\mathcal{X}^+$ denotes the polyhedral complex containing all (faces of) facets with positive weight.

Corollary 2.3.3

The map

$$\rho_{I|J} : \Big(\prod_{k\in I} \psi_k^{a_k} \cdot \mathcal{M}_{I\cup\{x\}}\Big) \times \Big(\prod_{k\in J} \psi_k^{a_k} \cdot \mathcal{M}_{J\cup\{y\}}\Big) \to (\varphi_{I|J} \cdot \psi_1^{a_1} \cdots \psi_n^{a_n} \cdot \mathcal{M}_n)^+$$

is a well-defined isomorphism of polyhedral complexes.

Proof. We have to check the conditions of definition 2.3.1. Using the lengths of the bounded edges as local coordinates on the cones and with the help of lemma 2.3.2, this is straightforward. The inverse map is given by splitting a given curve at its special vertex ν_{spec}. $\qquad\square$

2.3.2 The case of parameterized curves

For this subsection, we fix a reducible partition $I|J$ (i.e. $v_{I|J} = 0$). Let Δ_I, Δ_J be the corresponding splitting of the tropical degree Δ and let $A := I \cap [n]$, $B := J \cap [n]$ be the partition of the marked leaves. Let $Z = \max(x_1, y_1) \cdots \max(x_r, y_r) \cdot \mathbb{R}^r \times \mathbb{R}^r$ denote the diagonal in $\mathbb{R}^r \times \mathbb{R}^r$ and consider the map

$$\mathrm{ev}_x \times \mathrm{ev}_y : \mathcal{M}_{A\cup\{x\}}^{\mathrm{lab}}(\mathbb{R}^r, \Delta_I) \times \mathcal{M}_{B\cup\{y\}}^{\mathrm{lab}}(\mathbb{R}^r, \Delta_J) \to \mathbb{R}^r \times \mathbb{R}^r.$$

We define

$$Z_{I|J} := (\mathrm{ev}_x \times \mathrm{ev}_y)^*(Z).$$

Note that $|Z_{I|J}| \subseteq (\mathrm{ev}_x \times \mathrm{ev}_y)^{-1}(|Z|)$ holds, which implies $\mathrm{ev}_x|_{Z_{I|J}} = \mathrm{ev}_y|_{Z_{I|J}}$.

We furthermore define the morphism of polyhedral complexes $\pi_{I|J}$: $Z_{I|J} \to \mathcal{M}_n^{\text{lab}}(\mathbb{R}^r, \Delta)$ by

$$\mathcal{M}_{I \cup \{x\}} \times \mathbb{R}^r \times \mathcal{M}_{J \cup \{y\}} \times \mathbb{R}^r \to \mathcal{M}_{[n] \cup \Delta} \times \mathbb{R}^r$$
$$((p_I, P), (p_J, Q)) \mapsto (\rho(p_I, p_J), P),$$

where we choose the same anchor leaf for $\mathcal{M}_{I \cup \{x\}}^{\text{lab}}(\mathbb{R}^r, \Delta_I)$ and $\mathcal{M}_n^{\text{lab}}(\mathbb{R}^r, \Delta)$ and ρ is the glueing map for the abstract case defined in the previous subsection. So, as in the abstract case, $\pi_{I|J}$ glues two curves together at the extra leaves x and y — but now restricting to $Z_{I|J}$ makes sure that also the images of the curves in $\mathbb{R}^r$ can be glued together (without an ambiguity of the position of the glued curve).

Lemma 2.3.4

The map

$$\pi_{I|J} : \psi_1^{a_1} \cdots \psi_n^{a_n} \cdot Z_{I|J} \to (\varphi_{I|J} \cdot \psi_1^{a_1} \cdots \psi_n^{a_n} \cdot \mathcal{M}_n^{\text{lab}}(\mathbb{R}^r, \Delta))^+$$

is a well-defined isomorphism of polyhedral complexes.

Proof. This follows from the abstract case (cf. lemma 2.3.3) and from $\text{ev}_x|_{Z_{I|J}} = \text{ev}_y|_{Z_{I|J}}$. □

Remark 2.3.5

Obviously the positions of the marked leaves are preserved under $\pi_{I|J}$, i.e. (by abuse of notation) for $i \in I$ (resp. $j \in J$) the equation $\text{ev}_i \circ \pi_{I|J} = \text{ev}_i$ (resp. $\text{ev}_j \circ \pi_{I|J} = \text{ev}_j$) holds.

Lemma 2.3.6

Let $E = (\varphi_{I|J} \cdot \tau_{a_1}(C_1) \cdots \tau_{a_n}(C_n))_\Delta$ be a zero-dimensional cycle. Then all points of E lie in $(\varphi_{I|J} \cdot \psi_1^{a_1} \cdots \psi_n^{a_n} \cdot \mathcal{M}_n^{\text{lab}}(\mathbb{R}^r, \Delta))^+$.

Proof. By proposition 1.2.12 we can compute the weight of a point $p \in E$

locally around p in $X := \varphi_{I|J} \cdot \psi_1^{a_1} \cdots \psi_n^{a_n} \cdot \mathcal{M}_n^{\text{lab}}(\mathbb{R}^r, \Delta)$, namely we can focus on $\text{Star}_X(p)$. Assume $p \notin (\varphi_{I|J} \cdot \psi_1^{a_1} \cdots \psi_n^{a_n} \cdot \mathcal{M}_n^{\text{lab}}(\mathbb{R}^r, \Delta))^+$. Then curves corresponding to points in $\text{Star}_X(p)$ contain a bounded edge corresponding to the partition $I|J$ (see lemma 2.3.2). But as $I|J$ is chosen to be reducible, this edge is a contracted bounded edge whose length does not change the positions of the marked leaves in $\mathbb{R}^r$. Therefore, if we denote by $\text{ev} = \text{ev}_1 \times \ldots \times \text{ev}_n$ the product of all evaluation maps, the image of $\text{Star}_X(p)$ under ev has smaller dimension which implies $\text{ev}_*(\text{Star}_X(p)) = 0$. Hence, by the projection formula, the weight of p in E must be zero. $\square$

We now simplify the situation by choosing general incidence conditions. The following statement combines corollary 2.2.13, in particular item (c), and the preceding result.

Corollary 2.3.7

Let $E = (\varphi_{I|J} \cdot \tau_{a_1}(C_1) \cdots \tau_{a_n}(C_n))_\Delta$ be a zero-dimensional cycle. If we substitute the cycles C_i by general translations, we can assume that all points of E lie in the interior of a facet of $(\varphi_{I|J} \cdot \psi_1^{a_1} \cdots \psi_n^{a_n} \cdot \mathcal{M}_n^{\text{lab}}(\mathbb{R}^r, \Delta))^+$. This operation does not change the degree of E by remark 2.2.11.

Hence our provisional result can be formulated as follows.

Proposition 2.3.8

Let $I|J$ be a reducible partition and $E = (\varphi_{I|J} \cdot \tau_{a_1}(C_1) \cdots \tau_{a_n}(C_n))_\Delta$ be a zero-dimensional cycle. Then the equation

$$\langle \varphi_{I|J} \cdot \tau_{a_1}(C_1) \cdots \tau_{a_n}(C_n) \rangle_\Delta = \langle \tau_{a_1}(C_1) \cdots \tau_{a_n}(C_n) \cdot Z_{I|J} \rangle_{\Delta_I, \Delta_J}$$

holds.

Proof. We denote $X := \psi_1^{a_1} \cdots \psi_n^{a_n} \cdot Z_{I|J}$ and $Y := \varphi_{I|J} \cdot \psi_1^{a_1} \cdots \psi_n^{a_n} \cdot \mathcal{M}_n^{\text{lab}}(\mathbb{R}^r, \Delta)$ and assume that the conditions C_i are general. Then corollary 2.3.7 implies that, for each point $p \in E$, we have an isomorphism of cycles $\pi_{I|J} : \mathrm{Star}_X(\pi_{I|J}^{-1}(p)) \to \mathrm{Star}_Y(p)$. By the locality of the intersection product, it suffices to show that the weights of p and $\pi_{I|J}^{-1}(p)$ in their respective intersection products coincide. $\qquad\square$

2.3.3 Splitting the diagonal

Up to now, we have seen that intersecting with a "boundary" function $\varphi_{I|J}$ for reducible $I|J$ leads to intersection products in two smaller moduli spaces $\mathcal{M}_{A\cup\{x\}}^{\mathrm{lab}}(\mathbb{R}^r, \Delta_I)$ and $\mathcal{M}_{B\cup\{y\}}^{\mathrm{lab}}(\mathbb{R}^r, \Delta_J)$. However, the factor $(\mathrm{ev}_x \times \mathrm{ev}_y)^*(Z)$ still connects these two smaller spaces. In order to finally arrive at recursive equations of Gromov-Witten invariants, it is desirable to distribute this diagonal factor onto the two moduli spaces and to obtain independent intersection products there. In the algebro-geometric case, this can be easily done as the *class* of the diagonal Z in e.g. $\mathbb{P}^r \times \mathbb{P}^r$ can be written as the sum of products of classes in the factors (Künneth decomposition)

$$[Z] = [L^0 \times L^r] + [L^1 \times L^{r-1}] + \ldots + [L^r \times L^0],$$

where L^i denotes an i-dimensional linear space in $\mathbb{P}^r$. But this can *not* be imitated tropically: Our notion of rational equivalence is "too strong" for this application, as it is inspired by the idea that two rationally equivalent objects should be rationally equivalent in *any* toric compactification. Hence, by theorem 1.4.16 two rationally equivalent cycles in $\mathbb{R}^r \times \mathbb{R}^r$ must have the same outwards directions, which is surely not possible for a sum of cartesian products on the one hand and the diagonal on the other hand. However, we will discuss here how far the classical plan can be carried out anyways.

The general plan is the following: Set

$$X_I := (\tau_0(\mathbb{R}^r) \cdot \prod_{k\in I} \tau_{a_k}(C_k))_{\Delta_I} \quad \text{in } \mathcal{M}_{I\cup\{x\}}^{\mathrm{lab}}(\mathbb{R}^r, \Delta_I)$$

and

$$X_J := (\tau_0(\mathbb{R}^r) \cdot \prod_{k \in J} \tau_{a_k}(C_k))_{\Delta_J} \quad \text{in } \mathcal{M}^{\text{lab}}_{J \cup \{y\}}(\mathbb{R}^r, \Delta_J).$$

We want to compute the degree of

$$(\tau_{a_1}(C_1) \cdots \tau_{a_n}(C_n) \cdot Z_{I|J})_{\Delta_I, \Delta_J} = (\text{ev}_x \times \text{ev}_y)^*(Z) \cdot (X_I \times X_J),$$

or, by the projection formula,

$$\deg(Z \cdot (\text{ev}_x(X_I) \times \text{ev}_y(X_J))).$$

Now we would like to replace the diagonal Z by something like

$$S := \sum_{\alpha} (M_\alpha \times N_\alpha),$$

where M_α, N_α are cycles in $\mathbb{R}^r$ such that S intersects $\text{ev}_x(X_I) \times \text{ev}_y(X_J)$ like Z. As we cannot expect to find an S which is rationally equivalent to the diagonal, we need more information about what the push forwards $\text{ev}_x(X_I)$ and $\text{ev}_y(X_J)$ look like; in particular, we would like to know what their degree fans can look like. Let us formalize this first.

Let Ω be a complete unimodular fan in $\mathbb{R}^r$ and let $Z_k(\Omega)$ be the group of Ω-directional tropical fans X, i.e. $|X| \subseteq |\Omega^{(\dim(X))}|$. Fix a basis of $Z_*(\Omega) := \oplus_{k=0}^r Z_k(\Omega)$ denoted by $B_0, \ldots, B_m$ (where we may assume $B_0 = \{0\}$ and $B_m = \mathbb{R}^r$). More general, we call a tropical cycle X Ω-*directional* if the degree $\delta(X)$ is Ω-directional. For such a cycle there exist integer coefficients λ_e such that $X \sim \delta(X) = \sum_{e=1}^m \lambda_e B_e$.

Lemma 2.3.9

The linear map

$$Z_*(\Omega) \to \mathbb{Z}^{m+1},$$

$$X \mapsto (\deg(B_0 \cdot X), \ldots, \deg(B_m \cdot X)),$$

(where $\deg(.)$ *is set to be zero if the dimension of the argument is non-zero) is injective.*

Proof. Let $X \in Z_k(\Omega)$ be an element of the kernel, which implies that $\deg(X \cdot Y) = 0$ for all $Y \in Z_{r-k}(\Omega)$. Now we can use the proof of proposition 1.4.15. We just have to note that the functions φ_ϱ (which take value 1 on $u_{\varrho/\{0\}}$ and are zero on the other rays), when intersected with an Ω-directional cycle, obviously produce an Ω-directional cycle again. □

With respect to the basis $B_0, \ldots, B_m$ the map defined in the previous lemma has the matrix representation $\alpha := (\deg(B_e \cdot B_f))_{ef}$. Obviously α is a symmetric matrix. The lemma implies that this matrix is invertible (at least over $\mathbb{Q}$), and we denote the inverse by $(\beta_{ef})_{ef}$. The coefficients of this matrix can be used to replace the diagonal Z of $\mathbb{R}^r \times \mathbb{R}^r$ by a sum of products of cycles in the two factors (namely $\sum_{e,f} \beta_{ef}(B_e \times B_f)$) — at least with respect to Ω-directional cycles.

Lemma 2.3.10

Let $X \sim \sum_e \lambda_e B_e, Y \sim \sum_f \mu_e B_e$ *be two Ω-directional cycles in $\mathbb{R}^r$ with complementary dimension. Then*

$$\deg(Z \cdot (X \times Y)) = \deg(X \cdot Y) = \sum_{e,f} \deg(X \cdot B_e)\beta_{ef} \deg(Y \cdot B_f).$$

Proof. Denote $\lambda := (\lambda_1, \ldots, \lambda_m), \mu := (\mu_1, \ldots, \mu_m)$. We get

$$\sum_{e,f} \deg(X \cdot B_e)\beta_{ef} \deg(Y \cdot B_f) = (\alpha \cdot \lambda)^T \cdot \beta \cdot (\alpha \cdot \mu)$$

$$= \lambda^T \cdot \alpha^T \cdot \beta \cdot \alpha \cdot \mu$$
$$= \lambda^T \cdot \alpha \cdot \beta \cdot \alpha \cdot \mu$$
$$= \lambda^T \cdot \alpha \cdot \mu = \deg(X \cdot Y).$$

$\square$

Using this, our original goal of deriving a tropical splitting lemma can be formulated as follows.

Theorem 2.3.11 (Splitting lemma, cf. [Ko] 5.2.1)
Let $E = (\varphi_{I|J} \cdot \prod_{k=1}^n \tau_{a_k}(C_k))_{\Delta}^{\mathbb{R}^r}$ be a zero-dimensional cycle, where $I|J$ is a reducible partition. Moreover, let us assume that Ω is a complete unimodular fan such that (with the notations from above) $\mathrm{ev}_x(X_I)$ and $\mathrm{ev}_y(X_J)$ are Ω-directional. Let $B_0, \ldots, B_m$ be a basis of $Z_*(\Omega)$ and let $(\beta_{ef})_{ef}$ be the inverse matrix (over $\mathbb{Q}$) of $(\deg(B_e \cdot B_f))_{ef}$. Then the following equation holds:

$$\langle \varphi_{I|J} \cdot \prod_{k=1}^n \tau_{a_k}(C_k) \rangle_\Delta = \sum_{e,f} \langle \prod_{k \in I} \tau_{a_k}(C_k) \cdot \tau_0(B_e) \rangle_{\Delta_I} \beta_{ef} \langle \tau_0(B_f) \cdot \prod_{k \in J} \tau_{a_k}(C_k) \rangle_{\Delta_J}$$

Proof. The statement follows from the general plan above and proposition 2.3.8. $\square$

Remark 2.3.12
Of course, the basis $B_0, \ldots, B_m$ corresponds to a basis $\gamma_0, \ldots, \gamma_m$ of the cohomology groups of $\mathbf{X}(\Omega)$. As the cup-product and the intersection product of cycles are equivalent (cf. theorem 1.5.17), the corresponding matrix $(\deg(\gamma_e \cup \gamma_f))_{ef}$ is equal to α. This implies that the coefficients β_{ef} appearing in the tropical splitting lemma really are the same as in the

associated algebro-geometric version (cf. [Ko] 5.2.1).

2.3.4 The directions of families of curves

The above splitting lemma is only useful if, at least for a certain class of invariants, a fixed fan Ω exists such that all occurring push forwards $\mathrm{ev}_x(X_I)$ and $\mathrm{ev}_y(X_J)$ are Ω-directional. This is one of the main problems when transferring the algebro-geometric theory to the tropical set-up. However, in this subsection we show that in some cases the problem can be solved.

Remark 2.3.13

In the easiest case, namely if $r = 1$, the situation is trivial: There is one unique complete simplicial fan $\Omega = \{\mathbb{R}_{\leq 0}, \{0\}, \mathbb{R}_{\geq 0}\}$ and any subcycle is Ω-directional. Also, with $B_0 = \{0\}, B_1 = \mathbb{R}$, the statement of lemma 2.3.10 is obvious here.

Let us now consider curves in the plane, i.e. $r = 2$. Let $F = (\tau_0(\mathbb{R}^2) \cdot \prod_{k=1}^{n} \tau_{a_k}(C_k))_\Delta^{\mathbb{R}^2}$ be a one-dimensional family of plane curves (with unrestricted leaf x_0). We define $\Omega(F)$ to be the fan in $\mathbb{R}^2$ which contains all directions appearing in Δ and furthermore all rays in $\delta(C_k)$ if $\dim(C_k) = 1$ and $a_k > 0$.

Lemma 2.3.14

Let $F = (\tau_0(\mathbb{R}^2) \cdot \prod_{k=1}^{n} \tau_{a_k}(C_k))_\Delta^{\mathbb{R}^2}$ be a one-dimensional family of plane curves (with unrestricted leaf x_0). Let us furthermore assume that $a_k \leq 1$ if $\dim(C_k) = 2$ (i.e. if a leaf is not restricted by incidence conditions, at most one Psi-condition is allowed). Then $\mathrm{ev}_{0*}(F)$ is $\Omega(F)$-directional.

Proof. As before, we replace each factor $\psi_k^{a_k}$ by $\mathrm{ft}_0^*(\psi_k)^{a_k} + \mathrm{ft}_0^*(\psi_k)^{a_k-1} \cdot \varphi_{0,k}$ and multiply out. Consider the term without φ-factors: It is the fiber of $(\prod_{k=1}^{n} \tau_{a_k}(C_k))_\Delta$ (which is finite) under ft_0 (see family property of ft_0, proposition 2.2.9) and moreover the push forward of the fibre along ev_0

is just the sum/union of the images in $\mathbb{R}^r$ of the parameterized curves corresponding to the points in $(\prod_{k=1}^{n} \tau_{a_k}(C_k))_\Delta$. But these curves have degree Δ, thus by definition their images are $\Omega(F)$-directional.

So let us consider the term with the factor $\varphi_{0,k}$. Here, ev_0 and ev_k coincide (see lemma 2.2.19), so we can in fact compute the push forward along ev_k. As $\mathrm{ev}_k = \mathrm{ev}_k \circ \mathrm{ft}_0$ (by abuse of notation), we can first push forward along ft_0 and get the term $(\tau_{a_k-1}(C_k) \cdot \prod_{l \neq k} \tau_{a_l}(C_l))$.

Now, if $\dim(C_k) = 2$, by our assumptions $a_k - 1 \leq 0$; hence either we can use induction to prove the statement or this term does not appear at all.

On the other hand, if $\dim(C_k) = 0$ or 1, we can use the fact that the push forward is certainly contained in C_k — therefore, $\dim(C_k) = 0$ is trivial and $\dim(C_k) = 1$ works as we added the directions of C_k to $\Omega(F)$ if $a_k > 0$.

This finishes the proof, as all terms with more φ-factors vanish. $\qquad \square$

Remark 2.3.15
A weaker version of this lemma can be obtained by directly investigating on what the image under ev_0 of an unbounded ray in F looks like, using general conditions (see [MR08, lemma 3.7]).

Remark 2.3.16
Consider the family $F = (\tau_0(\mathbb{R}^2)\tau_0(P)\tau_2(\mathbb{R}^2))_1^{\mathbb{R}^2} = \mathrm{ev}_1^*(P) \cdot \psi_2^2 \cdot \mathcal{M}_3^{\mathrm{lab}}(\mathbb{R}^2, 1)$ of curves of projective degree 1. It consists of the following types of curves:

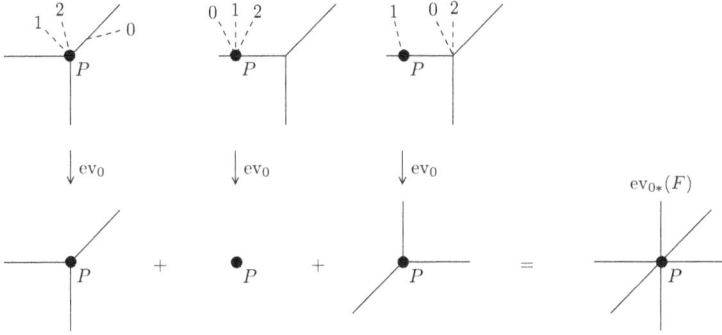

Its push forward along ev_0 also contains the inverted standard directions $(1,0)$, $(0,1)$ and $(-1,-1)$. Therefore this family is a counterexample of our statement if we drop the condition on the number of Psi-conditions allowed at leaves not restricted by incidence conditions.

Remark 2.3.17

For higher dimensions $(r > 2)$, only few cases are explored. However, for the important case of projective degree d and without any Psi-conditions, i.e. for a family $F = (\tau_0(\mathbb{R}^r) \cdot \prod_{k=1}^{n} \tau_0(C_k))_d$ of arbitrary dimension r, a proof exists that $ev_{0*}(F)$ is $\mathcal{L}_r^r$-directional (cf. [GZ]). We expect that a similar proof also works for Psi-conditions at point conditions. Beyond this, the behaviour of push forwards is mainly unknown up to now.

2.4 The WDVV equations and topological recursion

In this final section, we prove the tropical analogues of the WDVV and topological recursion equations — under certain restrictions. With the help of these equations, we show that certain tropical gravitational descendants coincide with their classical counterpart. There the computation of the classical invariants is reduced to counting tropical curves with certain valence and incidence conditions (cf. remark 2.2.14).

2.4.1 WDVV equations

Let x_i, x_j, x_k, x_l be pairwise different marked leaves and consider the forgetful map $\mathrm{ft} : \mathcal{M}_n^{\mathrm{lab}}(\mathbb{R}^r, \Delta) \to \mathcal{M}_{\{i,j,k,l\}}$.

Lemma 2.4.1

The equation

$$\mathrm{ft}^*(\varphi_{\{i,j\}|\{k,l\}}) = \sum_{\substack{I|J \\ i,j \in I, k,l \in J}} \varphi_{I|J}$$

holds, where the sum on the right hand side runs through all (also non-reducible) partitions with $i, j \in I$ and $k, l \in J$.

Proof. Note that $\mathrm{ft}(V_{I|J}) = V_{I\cap\{i,j,k,l\}|J\cap\{i,j,k,l\}}$. Therefore $\varphi(\mathrm{ft}(V_{I|J})) = 1$ if $i, j \in I, k, l \in J$ and zero otherwise, which proves the claim. $\square$

Now we face the crucial difference to the classical setting, due to the problems mentioned in remark 2.2.3: The right sum also runs over non-reducible partitions, which do not correspond to something in the algebro-geometric case. Let us add up only those $\varphi_{I|J}$ with $I|J$ non-reducible and denote the sum by $\phi_{i,j|k,l}$, i.e.

$$\phi_{i,j|k,l} := \sum_{\substack{I|J \text{ non-red.} \\ i,j \in I, k,l \in J}} \varphi_{I|J}.$$

We would like to show that $\phi_{i,j|k,l}$ is bounded, as then it does not change the degree of an intersection product and the term different from the classical setting vanishes. So let us investigate what this function measures:

Let $F = (\prod_{k=1}^n \tau_{a_k}(C_k))_\Delta$ be a one-dimensional family of curves with general conditions. Consider a facet σ of F representing curves *with* contracted bounded edge E (called *reducible curves*). Then we can change the length of E while keeping all other lengths and our curve still matches the incidence conditions. As our conditions are general, the set of curves fulfilling the incidence conditions set-theoretically is also one-dimensional (cf. remark 2.2.14). Hence, all curves in σ just differ by the length of E, whereas all other lengths are fixed. But this means that $\phi_{i,j|k,l}$ is constant on σ.

Now, let σ be a facet of F representing curves *without* contracted bounded edge E (called *non-reducible curves*). This means, for each reducible partition $I|J$, the respective function $\varphi_{I|J}$ is identically zero on σ. Therefore, on σ, $\phi_{i,j|k,l}$ coincides with $\mathrm{ft}^*(\varphi_{\{i,j\}|\{k,l\}})$.

Lemma 2.4.2
Let $F = (\prod_{k=1}^{n} \tau_{a_k}(C_k))_\Delta$ *be an one-dimensional family of curves with general conditions. Let* σ *be a facet of* F. *Then*

$$\phi_{i,j|k,l}|_\sigma = \begin{cases} \varphi_{\{i,j\}|\{k,l\}} \circ \mathrm{ft} & \text{if interior curves of } \sigma \text{ are non-reducible} \\ \text{const} & \text{otherwise.} \end{cases}$$

In other words: Proving that $\phi_{i,j|k,l}$ is bounded on a one-dimensional family F is equivalent to *proving that curves in F with large $\mathcal{M}_{i,j,k,l}$-coordinate must contain a contracted bounded edge*. This is the way of speaking in existing literature (e.g. [GM05, proposition 5.1], [KM06, proposition 6.1], [MR08, section 4]). We deal with this hard problem in subsection 2.4.3 and first state the desired results here.

Corollary 2.4.3 (cf. [Ko] 5.3.2)
Let $F = (\prod_{k=1}^{n} \tau_{a_k}(C_k))_\Delta$ *be a one-dimensional family of curves. Furthermore assume that* $\phi_{i,j|k,l}$ *is bounded. Then the equation*

$$\langle \mathrm{ft}^*(\varphi_{\{i,j\}|\{k,l\}}) \cdot \prod_{k=1}^{n} \tau_{a_k}(C_k) \rangle_\Delta = \sum_{\substack{I|J \text{ reducible} \\ i,j \in I, k,l \in J}} \langle \varphi_{I|J} \cdot \prod_{k=1}^{n} \tau_{a_k}(C_k) \rangle_\Delta$$

holds.

Proof. This follows from lemma 2.4.1 and lemma 1.4.4: If $\phi_{i,j|k,l}$ is bounded, the degree of

$$\langle \phi_{i,j|k,l} \cdot \prod_{k=1}^{n} \tau_{a_k}(C_k) \rangle_{\Delta}$$

is zero and hence this term can be omitted. $\qquad\square$

Remark 2.4.4

In classical Gromov-Witten theory, the WDVV equations are based on the crucial observation that $\overline{M}_{0,4}$ is isomorphic to $\mathbb{P}^1$ and therefore two points in $\overline{M}_{0,4}$ are rationally equivalent. Then, as in the tropical case, pulling back two "boundary" points of $\overline{M}_{0,4}$ along the forgetful morphism ft : $\overline{M}_{0,n}(\mathbf{X}, \beta) \to \overline{M}_{0,4}$ leads to relations between the irreducible boundary divisors of $\overline{M}_{0,n}(\mathbf{X}, \beta)$ (assume that $\mathbf{X}$ is projective homogeneous again). On the tropical side, this observation is replaced by the fact that $\mathcal{M}_4$ is isomorphic to L_1^2 and that two boundary functions on $\mathcal{M}_4$, say $\phi_{i,j|k,l}$ and $\phi_{i,k|j,l}$ only differ in a linear function. Hence they even define the same Cartier divisor (and their Weil divisor equals the single vertex in $\mathcal{M}_4$ in both cases). Hence the pulled back Cartier divisors also coincide and this provides relations between the boundary divisors in the same way as classically.

We can now state the following version of the WDVV equations:

As before, we fix a complete unimodular fan Ω and a basis $B_0, \ldots, B_m$ of $Z_*(\Omega)$. Furthermore, let $(\beta_{ef})_{ef}$ be the inverse matrix (over $\mathbb{Q}$) of the matrix $(\deg(B_e \cdot B_f))_{ef}$.

Theorem 2.4.5 (WDVV equations, cf. [Ko] 5.3.3)

Let $F = (\prod_{k=1}^{n} \tau_{a_k}(C_k))_{\Delta}$ be a one-dimensional family of curves and fix four pairwise different marked leaves x_i, x_j, x_k, x_l. Moreover, we assume that the following conditions hold:

(a) *For any reducible partition $I|J$ with $i,j \in I; k,l \in J$ or $i,k \in I; j,l \in J$ the push forwards $\mathrm{ev}_x(X_I)$ and $\mathrm{ev}_y(X_J)$ are Ω-directional (with notations from section 2.3).*

(b) *The functions $\phi_{i,j|k,l}$ and $\phi_{i,k|j,l}$ are bounded on F.*

Then the WDVV equation

$$
\sum_{\substack{I|J \text{ reducible} \\ i,j \in I, k,l \in J}} \sum_{e,f} \langle \prod_{k \in I} \tau_{a_k}(C_k) \cdot \tau_0(B_e) \rangle_{\Delta_I} \beta_{ef} \langle \tau_0(B_f) \cdot \prod_{k \in J} \tau_{a_k}(C_k) \rangle_{\Delta_J}
$$
$$
= \sum_{\substack{I|J \text{ reducible} \\ i,k \in I, j,l \in J}} \sum_{e,f} \langle \prod_{k \in I} \tau_{a_k}(C_k) \cdot \tau_0(B_e) \rangle_{\Delta_I} \beta_{ef} \langle \tau_0(B_f) \cdot \prod_{k \in J} \tau_{a_k}(C_k) \rangle_{\Delta_J}
$$

holds, where the sums run through reducible partitions only.

Proof. The statement follows from the splitting lemma 2.3.11, corollary 2.4.3 and the fact that on $\mathcal{M}_{\{i,j,k,l\}}$ the functions $\varphi_{\{i,j\}|\{k,l\}}$ and $\varphi_{\{i,k\}|\{j,l\}}$ define the same Cartier divisor (cf. remark 2.4.4). $\qquad\square$

Remark 2.4.6 (Unlabelled degrees)
In the algebro-geometric version of these equations (cf. [Ko, 5.3.3] or [FP95, equations (54) and (55)]) the big sum(s) usually run like $\sum_{\beta_1,\beta_2}$ $\sum_{A,B}$, where β_1, β_2 are cohomology classes such that $\beta_1 + \beta_2 = \beta$ and $A \cup B = [n]$ is a partition of the marks. We can proceed accordingly and let our sums run through unlabelled instead of labelled degrees. If we collect all reducible partitions $I \cup J = \Delta \cup [n]$ such that the unlabelled degrees $\delta(\Delta_I)$ and $\delta(\Delta_J)$ coincide, we get precisely $\frac{\Delta!}{\Delta_I! \cdot \Delta_J!}$ elements. But then, as mentioned at the beginning of section 2.2, counting curves with labelled non-contracted leaves leads to an overcounting by the factor $\Delta!$ (modulo automorphisms), i.e. if $\delta := \delta(\Delta)$ is a positive one-dimensional

tropical fan, we define

$$\langle \prod_{k=1}^{n} \tau_{a_k}(C_k) \rangle_{\delta} := \frac{1}{\Delta!} \langle \prod_{k=1}^{n} \tau_{a_k}(C_k) \rangle_{\Delta}.$$

Hence, when switching to "unlabelled" invariants, the above factor $\frac{\Delta!}{\Delta_I! \cdot \Delta_J!}$ cancels and we obtain the equation

$$\sum_{\substack{\delta_I, \delta_J \\ \delta_I + \delta_J = \delta}} \sum_{\substack{A \cup B = [n] \\ i,j \in A, k,l \in B}} \sum_{e,f} \langle \prod_{k \in A} \tau_{a_k}(C_k) \cdot \tau_0(B_e) \rangle_{\delta_I} \beta_{ef} \langle \tau_0(B_f) \cdot \prod_{k \in B} \tau_{a_k}(C_k) \rangle_{\delta_J}$$

$$= \sum_{\substack{\delta_I, \delta_J \\ \delta_I + \delta_J = \delta}} \sum_{\substack{A \cup B = [n] \\ i,k \in A, j,l \in B}} \sum_{e,f} \langle \prod_{k \in A} \tau_{a_k}(C_k) \cdot \tau_0(B_e) \rangle_{\delta_I} \beta_{ef} \langle \tau_0(B_f) \cdot \prod_{k \in B} \tau_{a_k}(C_k) \rangle_{\delta_J}.$$

It is another issue if, in the respective classical equation, all cohomology sums $\beta_1 + \beta_2 = \beta$ which are "positive enough" such that their corresponding term in the sum does not vanish, correspond to positive tropical cycles represented by tropical degrees. This will be addressed to remark 2.4.19.

2.4.2 Topological recursion

In the same flavour as in the previous subsection, we also formulate a tropical version of the equations known as "topological recursion".

Let x_i, x_k, x_l be pairwise different marked leaves. We know from lemma 2.1.26 that we can express the Psi-divisor ψ_i in terms of "boundary" divisors, namely

$$\operatorname{div}(\psi_i) = \sum_{\substack{I|J \\ i \in I, k,l \in J}} \operatorname{div}(\varphi_{I|J}).$$

Again we give a name to the term that has no algebro-geometric counterpart,

$$\phi_{i|k,l} = \sum_{\substack{I|J \text{ non-red.} \\ i \in I; k,l \in J}} \varphi_{I|J}.$$

As in the previous subsection, we can describe this function as follows.

Lemma 2.4.7

Let $F = (\prod_{k=1}^{n} \tau_{a_k}(C_k))_\Delta$ be a one-dimensional family of curves with general conditions. Let σ be a facet of F. Then

$$\phi_{i|k,l}|_\sigma = \begin{cases} \sum \text{lengths of edges that separate } i \text{ from } k,l \\ \qquad \text{if interior curves of } \sigma \text{ are non-reducible,} \\ \text{constant} \\ \qquad \text{otherwise.} \end{cases}$$

Again, we fix a complete unimodular fan Ω and a basis $B_0, \ldots, B_m$ of $Z_*(\Omega)$. Furthermore, let $(\beta_{ef})_{ef}$ be the inverse matrix (over $\mathbb{Q}$) of the matrix $(\deg(B_e \cdot B_f))_{ef}$.

Theorem 2.4.8 (Topological recursion, cf. [Ko] 5.4.1)

Let $F = (\prod_{k=1}^{n} \tau_{a_k}(C_k))_\Delta$ be a one-dimensional family of curves and fix three pairwise different marked leaves x_i, x_k, x_l. Moreover, we assume that the following conditions hold:

(a) *For any reducible partition $I|J$ with $i \in I; k, l \in J$ the push forwards $\text{ev}_x(X_I)$ and $\text{ev}_y(X_J)$ are Ω-directional (with notations from section 2.3).*

(b) *The function $\phi_{i|k,l}$ is bounded on F.*

Then the topological recursion

$$\langle \psi_i \cdot \prod_{k=1}^{n} \tau_{a_k}(C_k) \rangle_\Delta$$

$$= \sum_{\substack{I|J \text{ reducible} \\ i \in I, k, l \in J}} \sum_{e,f} \langle \prod_{k \in I} \tau_{a_k}(C_k) \cdot \tau_0(B_e) \rangle_{\Delta_I} \beta_{ef} \langle \tau_0(B_f) \cdot \prod_{k \in J} \tau_{a_k}(C_k) \rangle_{\Delta_J}$$

holds, where the sum runs through reducible partitions only.

Proof. As in the classical case, we replace ψ_i by a sum of boundary divisors

$$\psi_i = \sum_{\substack{I|J \\ i \in I, k, l \in J}} \operatorname{div}(\varphi_{I|J}).$$

As the term

$$\langle \phi_{i|k,l} \cdot \prod_{k=1}^{n} \tau_{a_k}(C_k) \rangle_\Delta$$

vanishes, the splitting lemma 2.3.11 proves the claim. $\square$

Remark 2.4.9 (Unlabelled degrees)

In the same way as in remark 2.4.6, we obtain the "unlabelled" version

$$\langle \psi_i \cdot \prod_{k=1}^{n} \tau_{a_k}(C_k) \rangle_\delta =$$

$$\sum_{\substack{\delta_I, \delta_J \\ \delta_I + \delta_J = \delta}} \sum_{\substack{A \cup B = [n] \\ i \in A, k, l \in B}} \sum_{e,f} \langle \prod_{k \in A} \tau_{a_k}(C_k) \cdot \tau_0(B_e) \rangle_{\delta_I} \beta_{ef} \langle \tau_0(B_f) \cdot \prod_{k \in B} \tau_{a_k}(C_k) \rangle_{\delta_J},$$

where $\delta, \delta_I, \delta_J$ denote unlabelled degrees, i.e. positive one-dimensional tropical fans.

2.4.3 Contracted bounded edges

It is certainly unsatisfactory to finish this thesis with two theorems whose list of assumptions is as long as the list of results. Therefore, the goal of the rest of this thesis is to verify these assumptions for plane curves of certain degrees and to use the theorems in this particular case. Recall that subsection 2.3.4 was devoted to the study of assumption (a) (of the previous theorems) — lemma 2.3.14 is sufficient for our purposes. This subsection here deals with assumption (b). The material is an extension of the "contracted edge" argument in [GM05]. In particular, the proof of lemma 2.4.14 is essentially contained in the proof of [GM05, proposition 5.1].

As a preparation for the more difficult case of plane curves, we first assume $r = 1$. We often use the following notation: Let τ be an edge of a parameterized curve and let $\nu \in \tau$ be a vertex, then the *direction vector v of τ* is the image of $u_{\tau/\nu}$ under the linear part of h. Moreover, for the sake of simplicity we often denote an edge by the same letter v as its direction vector (the vertex is clear from the context).

Lemma 2.4.10

Let $P_1, \ldots, P_n$ be points in general position in $\mathbb{R}^1$ and let $F = (\prod_{k=1}^{n} \tau_{a_k}(P_k))_d^{\mathbb{R}^1}$ be a one-dimensional family in $\mathcal{M}_n^{\mathrm{lab}}(\mathbb{R}^1, d)$. Then for any choice of marked leaves x_i, x_j, x_k, x_l, the functions $\phi_{i,j|k,l}$ and $\phi_{i|k,l}$ are bounded on F.

Proof. For general conditions, F set-theoretically coincides with the set of curves satisfying the given incidence and valence conditions (cf. remark 2.2.14). Consider a general curve $C \in F$. Then C is also a general curve in the Psi-product $X := \prod_{k=1}^{n} \psi_k^{a_k}$. As we cut down X by n point conditions and $\dim(F) = 1$, the dimension of X must be $n + 1$, hence C contains n bounded edges. This implies that C, as it is a rational curve, has $n + 1$ vertices. But all marked leaves $x_k, k \in [n]$ lie at different vertices, due to the general position of the points P_i. Therefore there exists a vertex ν not adjacent to a marked leaf $x_k, k \in [n]$. Now, one of the three edges adjacent to V might be a contracted bounded edge. Then the deformation of C in F is given by changing the length of this edge, but this does not affect $\phi_{i,j|k,l}$ and $\phi_{i|k,l}$ by definition. Otherwise, if all of the adjacent edges are non-contracted, the one-dimensional deformation of C in F is given by moving ν (and changing the lengths accordingly). The picture looks like this:

Note that the edge v cannot be unbounded as its direction "vector" $v = -v_1 - v_2$ is not primitive. Therefore, if this deformation is supposed to be unbounded, v_1, v_2 must be unbounded. But in this case only the length of v grows infinitely. But as v does not separate any marked leaves, this does not affect $\phi_{i,j|k,l}$ and $\phi_{i|k,l}$. $\qquad \square$

Now let us consider the case of plane curves, i.e. $r = 2$. We fix the following notation: Let $F = (\prod_{k=1}^{n} \tau_{a_k}(C_k))_{\Delta}^{\mathbb{R}^2}$ be a one-dimensional family of plane curves with general conditions and let $L \cup M \cup N = [n]$ be the partition of the labels such that

$$\mathrm{codim}(C_k) = \begin{cases} 0 & \text{if } k \in L, \\ 1 & \text{if } k \in M, \\ 2 & \text{if } k \in N. \end{cases}$$

First we study how the deformation of a general curve C in F can look like.

Lemma 2.4.11 (Variation of [MR08] 4.4)

Let us assume

i) $a_k = 0$ for all $k \in L \cup M$, i.e. Psi-conditions are only allowed together *with point conditions.*

Then the following holds:

Let σ be a facet of F and let $C \in \sigma$ be a general curve. Then the deformation of C inside σ is described by one of the following cases:

(I) *C contains a* contracted bounded edge. *Then the deformation inside σ is given by changing the length of this edge arbitrarily.*

(II) *C has a 3-valent* degenerated vertex ν *of one of the following three types:*

  a) *One of the adjacent edges is a marked leaf $i \in L$.*

  b) *One of the adjacent edges is a marked leaf $j \in M$ and the linear spans of the corresponding cycle C_j at $\mathrm{ev}_j(C)$ and of the other two edges adjacent to ν coincide (i.e. the curves C and the cycle C_j do* not *intersect transversally at $\mathrm{ev}_j(C)$).*

  c) *All edges adjacent to V are non-contracted, but their span near ν is still only one-dimensional; w.l.o.g. we denote the isolated edge of ν by v and the two edges on the other side by v_1, v_2.*

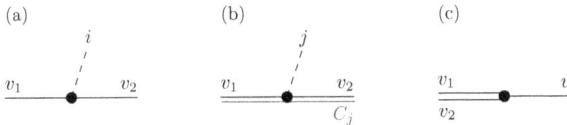

In all these cases the deformation inside σ is given by moving V.

(III) *C contains a* movable string S, *i.e. a 2-valent subgraph of C homeomorphic to $\mathbb{R}$ such that all edges are non-contracted and all vertices of S are 3-valent in C and not degenerated in the sense of case (II). Then the deformation of C is given by moving S while all vertices not contained in S remain fixed (in particular, only edges in or adjacent to S change their lengths).*

Proof. Again, for general conditions, F set-theoretically coincides with the set of curves satisfying the given incidence and valence conditions. Thus finding the deformation of C inside σ is the same as finding a way of changing the position and the length of the bounded edges of C such that the resulting curve still meets the incidence conditions C_k.

It is obvious that in the cases (I) and (II) changing the length of the contracted bounded edge respectively moving the degenerated vertex ν leads to such deformations.

In case (III) the non-degeneracy of the vertices makes sure that both leaves of S are non-contracted and that a small movement of one of these leaves leads to a well-defined movement of the whole string: Take one of the leaves of the string and move it slightly in a non-zero direction modulo its linear span. Consider the next vertex ν and let v be the adjacent edge not contained in the string. Then two things can happen:

A: If v is non-contracted , our moved leaf will meet the affine span of v at some point P (as ν is non-degenerated). So we change the length of v such that it ends at P (while keeping the position of its second vertex fixed). Then we also move the second edge of the string to P and go on to the next vertex.

B: If v is contracted , our assumptions ensure that it is a marked leaf $j \in M$ and that the corresponding cycle C_j intersects our curve transversally at ν. Thus our moved edge still meets C_j at some point and by changing the lengths of the adjacent edges appropriately, the obtained curve still meets C_j.

In this way we can make our way through the string and finally obtain a deformation of the whole curve. Note that the non-degeneracy of all the vertices ensures that all edges of the string *must* change their positions modulo their linear span and, hence, that all edges adjacent to, but not contained in the string *must* change their lengths. In particular this means that we cannot have more non-contracted leaves adjacent to our string: Then we would have two different strings providing two independent deformations of the curves inside σ, which is a contradiction as σ is one-dimensional.

Let us summarize: Our string S is generated by two unique non-contracted leaves i_1, i_2, all of its vertices are 3-valent and the adjacent edges not contained in the string are either bounded edges or marked leaves in M, where the corresponding line C_j intersects transversally. T he deformation only moves the string S; the adjacent edges are shortened or lengthened and the other parts of the curve remain fixed.

Finally, this list of cases is really complete, as C always contains a string whose vertices are 3-valent in C and whose leaves are either non-contracted leaves or marked leaves in L. This results from the following computation: We know $\dim(F) = 1$, $\mathrm{codim}(F) = \#M + 2\#N + \sum_{k \in N} a_k$ and $\dim(\mathcal{M}_n^{\mathrm{lab}}(\mathbb{R}^\Delta,)) = \#L + \#M + \#N + \#\Delta - 3 + 2$. Plugging in all this in $\dim(F) + \mathrm{codim}(F) = \dim(\mathcal{M}_n^{\mathrm{lab}}(\mathbb{R}^\Delta,))$ leads to

$$\#L + \#\Delta = \#N + \sum_{k \in N} a_k + 2.$$

On the other hand we can compute the number of connected components of $C \setminus \bigcup_{k \in N} \bar{x}_k$ (i.e. we remove all marked leaves x_k with point conditions, together with the adjacent vertex). Removing $\bar{x}_k$ increases the number of connected components by $a_k + 1$ as the valence of the adjacent vertex is $a_k + 3$. So, after removing all leaves in N, we arrive at $1 + \#N + \sum_{k \in N} a_k$ connected components. The above equation tells us that there is one more leaf in $L \cup \Delta$ than there are connected components and therefore at least two leaves $i_1, i_2 \in L \cup \Delta$ lie in the same component. But then, if the string between i_1 and i_2 does not satisfy the assumptions of case (III), one of the other cases applies. $\qquad\square$

Now we know how a general curve $C \in F$ can be deformed. In a second step, we will now focus on unbounded deformations.

Definition 2.4.12

A fan Ω in $\mathbb{R}^2$ is called *strongly unimodular* if *any* two independent primitive vectors generating rays of Ω form a basis of $\mathbb{Z}^2$.

For a given degree Δ let $\Omega(\Delta)$ be the fan consisting of all rays generated by a direction vector appearing in Δ (i.e. $\Omega(\Delta)$ is the fan supporting $\delta(\Delta)$). A degree Δ in $\mathbb{R}^2$ is called *strongly unimodular* if $\Omega(\Delta)$ is strongly unimodular and if all direction vectors appearing in Δ are primitive. This ensures that for every pair of independent vectors v_1, v_2 appearing in Δ, the dual triangle to the fan spanned by v_1, v_2 and $-(v_1 + v_2)$ does not contain lattice points besides its vertices.

Remark 2.4.13

It is easy to check that the rays of a strongly unimodular fan can be identified via lattice isomorphisms to some rays of the fan $\Omega_{\mathcal{B}l_3(\mathbb{P}^2)}$

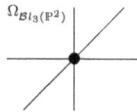

associated to the blow up $\mathcal{B}l_3(\mathbb{P}^2)$ of the three torus-fixed points of $\mathbb{P}^2$. The following is a complete list of such fans (up to isomorphisms).

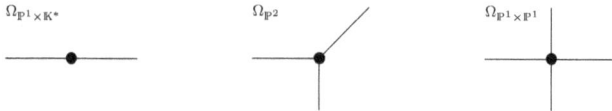

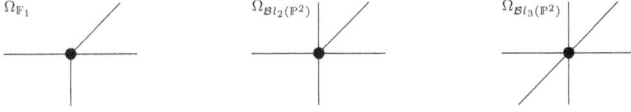

Lemma 2.4.14 (Variation of [MR08] 4.4)

We assume

i) $a_k = 0$ *for all* $k \in L \cup M$,

ii) Δ *is strongly unimodular.*

Then the following holds:

Let σ be an unbounded *facet of F and let $C \in \sigma$ be a general curve. Then the deformation of C in σ is described by one of the following cases:*

(I) *C contains a contracted bounded edge whose length can be changed arbitrarily.*

(II) *C has a 3-valent degenerated vertex ν of one the three types described above. Furthermore, in the cases (a) and (b) (of 2.4.11 (II)) one of the edges v_1, v_2 is bounded, the other one unbounded, whereas in case (c) the edge v is bounded and v_1, v_2 are unbounded.*

(III) *C contains a movable string S with two non-contracted leaves v_1, v_2 and only one adjacent bounded edge w. The deformation of C is given by increasing the length of w.*

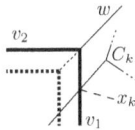

Furthermore, if $x_k, k \in M$ is a marked leaf adjacent to S, then $h(x_k)$ is a general point in an unbounded facet of C_k whose outgoing direction vector v lies in the interior of the cone spanned by v_1, v_2.

Proof. Nothing new happens in the cases (I), (II) (a) and (b). In case (II) (c), the only claim is that the edge v cannot be unbounded as $v = -v_1 - v_2$ is not primitive. Therefore the two edges on the other side of V must be unbounded.

So let us consider case (III), i.e. assume that the deformation of σ is given by a movable string S, i.e. a 2-valent subgraph of C homeomorphic to $\mathbb{R}$ such that all edges are non-contracted and all vertices of S are 3-valent in C and not degenerated. Note that in the first part of this argument as well as in the following picture, marked leaves $x_k, k \in M$ adjacent to S do not matter and are therefore omitted. In the following, we follow the respective proofs in [MR08] resp. [GM05].

If there are bounded edges adjacent to both sides of S as in picture (a) below then the movement of the string is bounded to both sides. So we only have to consider the case when all adjacent bounded edges of S are on the same side of S, say on the right hand side as in picture (b) below. We label the edges of S (respectively, their direction vectors) by $v_1, \ldots, v_k$ and the adjacent bounded edges of the curve by $w_1, \ldots, w_{k-1}$ as in the picture. As above the movement of the string to the right is bounded. If one of the directions w_{i+1} is obtained from w_i by a left turn (as it is the case for $i = 1$ in the picture) then the edges w_i and w_{i+1} meet on the left of S. This restricts the movement of the string to the left, too, since the corresponding edge v_{i+1} then shrinks to length 0.

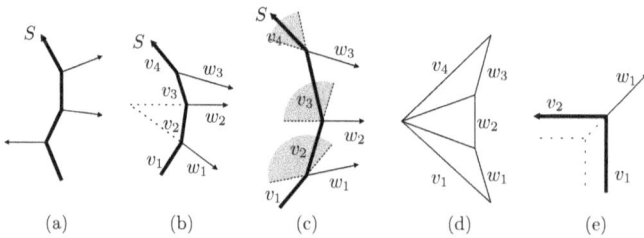

(a)   (b)   (c)   (d)   (e)

So we can assume that for all i the direction w_{i+1} is either the same as w_i or obtained from w_i by a right turn as in picture (c). The balancing condition then shows that for all i both the directions v_{i+1} and $-w_{i+1}$ lie in the angle between v_i and $-w_i$ (shaded in the picture above). Therefore, all directions v_i and $-w_i$ lie within the angle between v_1 and $-w_1$. In particular, the image of the string S cannot have any self-intersections in $\mathbb{R}^2$. We can therefore pass to the (local) dual picture (d) where the edges dual to w_i correspond to a concave side of the polygon whose other two edges are the ones dual to v_1 and v_k.

But note that there are no such concave polygons *with integer vertices*, as the two outer edges are dual to v_1 and v_k which form a lattice basis by definition of the strongly unimodular degree. Therefore the string is connected to the rest of the curve by exactly one bounded edge w_1, as stated. The remaining part of the claim follows directly. $\qquad\square$

Theorem 2.4.15

Let x_i, x_j, x_k, x_l be pairwise different marked leaves and let us assume

i) $a_k = 0$ for all $k \in L \cup M$,

ii) Δ is strongly unimodular,

iii) if $i, j \in M$ (resp. $k, l \in M$), then for any pair of independent direction vectors v_1, v_2 appearing in Δ, the interior of the cone spanned by v_1, v_2 does not intersect both degrees $\delta(C_i)$ and $\delta(C_j)$ (resp. $\delta(C_k)$ and $\delta(C_l)$).

Then $\phi_{i,j|k,l}$ is bounded.
If we additionally require

iv) $i \in N$,

then also $\phi_{i|k,l}$ is bounded.

Proof. As conditions i) and ii) hold, we can apply lemma 2.4.14, which describes the unbounded facets of F. We have to show that $\phi_{i,j|k,l}$ (resp.

$\phi_{i|k,l}$) is bounded on these facets. In case (I), the only length changing is that of a contracted edge and therefore not measured by both $\phi_{i,j|k,l}$ and $\phi_{i|k,l}$. In case (II), the edge whose length is growing infinitely cannot separate more then one marked leaf $x_k, k \in L \cup M$ from the others. Therefore this length cannot contribute to $\phi_{i,j|k,l}$ and — by condition iv) — to $\phi_{i|k,l}$. Finally, condition iii) (and also condition iv)) is chosen such that $\phi_{i,j|k,l}$ and $\phi_{i|k,l}$ are also bounded in case (III). Namely, if for example x_i and x_j, $i, j \in M$, are adjacent to the string S, then both $\delta(C_i)$ and $\delta(C_j)$ must contain a ray which lies in the interior of the cone spanned by v_1, v_2, which contradicts iii). $\qquad \square$

Remark 2.4.16

The conditions i) – iv) appearing in the above statements are not only sufficient but, in most cases, also necessary for the statements to hold:

iv) If condition iv) in theorem 2.4.15 is not satisfied, we can get the following things:

- If $i \in L$, then the degenerated vertex of type (a) leads to an unbounded $\phi_{i|k,l}$.

- If $i \in M$ and ρ is a ray in C_i whose direction vector v_ρ also appears in Δ, then in general we will find curves in F with a degenerated vertex of type (b), whose unbounded movement will make $\phi_{i|k,l}$ unbounded.

- If $i \in M$ and ρ is a ray in C_i whose direction vector v_ρ lies between two direction vectors v_1, v_2 appearing in Δ, in general this leads to curves in F with unbounded deformations of case (III) such that the outward directions are v_1, v_2 and such that x_i is adjacent to the moved string. So again, $\phi_{i|k,l}$ is in general unbounded.

iii) If condition iii) is not satisfied, we will in general get unbounded deformations of the following type:

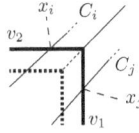

In this case we have $i, j \in M$ and the interior of the cone spanned by v_1, v_2 contains direction vectors of both C_i and C_j. As in general x_k, x_l lies on the other side of the growing edge w, $\phi_{i,j|k,l}$ is unbounded.

ii) If we drop condition ii), i.e. if we allow non-unimodular degrees Δ, two things can happen: If we allow non-primitive direction vectors, then we get deformations of type (II) (c) with unbounded edge v. Therefore the lengths of v_1 and v_2, which can in general separate arbitrary marked leaves, grow infinitely. If we drop the condition that $\Omega(\Delta)$ is strongly unimodular, then the description of unbounded deformations of case (III) in 2.4.14 becomes incorrect, as there appear more complicated strings with more adjacent bounded edges than just one. The example of the fan associated to the second Hirzebruch surface $\mathbb{F}_2$ is analyzed in detail in [Fra08] and [FM08, e.g. 2.10].

i) If we drop condition i), i.e. if we allow Psi-conditions also at marked leaves which are not fixed by points, we end up with more complicated kinds of deformations of general curves in F. The following picture shows an example of an unbounded deformation in a one-dimensional family of plane curves of projective degree 2. Here, C has to meet all the four tropical lines $C_1, \ldots, C_4$ with one Psi-condition. Note that the indicated deformation of C is indeed unbounded and that the length of the $(1, -1)$-edge e grows infinitely.

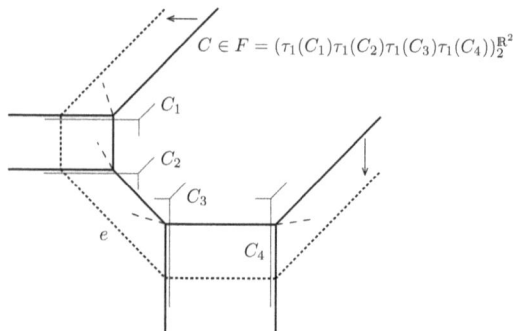

This example can be extended in the following way: One can glue arbitrary curves (fixed by appropriate conditions) to the non-contracted leaves of C in direction $(1,1)$, obtaining more families admitting such a deformation. In particular, the edge e can separate arbitrary kinds of points, showing that in general $\phi_{i,j|k,l}$ and $\phi_{i|k,l}$ can be unbounded for any choice of i, j, k, l.

For higher dimensions, let us mention the following important case where an affirmative result is known as well.

Theorem 2.4.17 ([Zim07] 4.86)
Let $F = (\prod_{k=1}^{n} \tau_0(V_k))_d^{\mathbb{R}^r}$ be a one-dimensional family of curves of projective degree d in $\mathbb{R}^r$ which do not satisfy Psi-conditions, but incidence conditions given by classically linear spaces $V_k \subseteq \mathbb{R}^r$. Then for any choice of $\{i, j, k, l\} \in [n]$ the function $\phi_{i,j|k,l}$ is bounded on F.

2.4.4 Comparison to the classical gravitational descendants

In the special case of an empty degree, denoted by $\Delta = 0$, the situation is analogous to the algebro-geometric one.

Lemma 2.4.18 (The degree zero case)

Let $Z = (\prod_{k=1}^{n} \tau_{a_k}(C_k))_0$ be a zero-dimensional intersection product in $\mathcal{M}_n^{\mathrm{lab}}(\mathbb{R}^r, 0)$. Then $\deg(Z)$ is non-zero if and only if $\sum_{k=1}^{n} \mathrm{codim}(C_k) = r$ (or equivalently $\sum_{k=1}^{n} a_k = n - 3$). In this case,

$$\deg(Z) = \binom{n-3}{a_1, \ldots, a_n} \deg(C_1 \cdots C_k)$$

holds.

Proof. By definition $\mathcal{M}_n^{\mathrm{lab}}(\mathbb{R}^r, 0)$ is isomorphic to $\mathcal{M}_n \times \mathbb{R}^r$. Moreover, as $\Delta = 0$, all evaluation maps ev_i coincide with the projection onto the second factor, which we therefore denote by ev. Now let $X := \prod_{k=1}^{n} \psi_k^{a_k} = (\prod_{k=1}^{n} (\psi_k^{\mathrm{abstr}})^{a_k}) \times \mathbb{R}^r$ be the intersection of all Psi-divisors. Then the projection formula applied to ev yields

$$\deg(Z) = \deg(C_1 \cdots C_n \cdot \mathrm{ev}_*(X)).$$

But $\mathrm{ev}_*(X)$ is non-zero if and only if $\sum_{k=1}^{n} a_k = n - 3$. If so, by remark 2.1.24 we know $\mathrm{ev}_*(X) = \binom{n-3}{a_1, \ldots, a_n} \cdot \mathbb{R}^r$, which proves the statement. $\quad\square$

Remark 2.4.19

The goal of the following theorem is to show that certain tropical and classical gravitational descendants coincide. The idea is to show that — under the restrictions which we accumulated in the preceding sections — both sets of numbers satisfy the same WDVV and topological recursion equations, which are sufficient to determine the numbers from some initial values. However, there is one further problem concerning this plan, which we already mentioned in remark 2.4.6. The classical WDVV and topological recursion equations run through splittings of the given cohomology class β into sums $\beta = \beta_1 + \beta_2$. As $\overline{M}_{0,n}(\mathbf{X}, \beta)$ is empty if β is not effective, we can restrict to effective classes β, β_1, β_2.

Now, for $\mathbb{P}^2$ and $\mathbb{P}^1 \times \mathbb{P}^1$, effectivity is equivalent to the fact that the associated one-dimensional tropical fans are positive (as $\mathbb{P}^2$ and $\mathbb{P}^1 \times \mathbb{P}^1$ do not contain curves with negative self-intersection). So a splitting $\beta = \beta_1 + \beta_2$ of effective cohomology classes corresponds bijectively to a sum of unlabelled tropical degrees $\delta = \delta_1 + \delta_2$, and therefore the tropical and classical equations are really equivalent in this case.

However, for the blow ups of $\mathbb{P}^2$ in up to three torus-fixed points (i.e. for $\mathbb{F}_2$, $\mathcal{B}l_2(\mathbb{P}^2)$ and $\mathcal{B}l_3(\mathbb{P}^2)$, cf. remark 2.4.13), the same argument fails as the exceptional divisors induce tropical fans with negative weights. The following picture shows the example of the tropical fan associated to the exceptional divisor $V(\varrho)$ of $\mathbb{F}_1$.

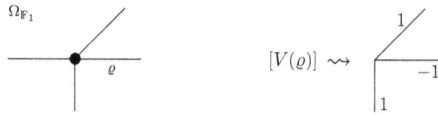

$$\Omega_{\mathbb{F}_1} \qquad\qquad [V(\varrho)] \rightsquigarrow \qquad \begin{array}{c} 1 \\ -1 \\ 1 \end{array}$$

Up to now, it is unknown if the following theorem can be extended to these toric varieties (possibly by adding suitable correction terms as in [FM08]). This needs to be addressed in further work. Here, we restrict to $\mathbb{P}^2$ and $\mathbb{P}^1 \times \mathbb{P}^1$ for simplicity and obtain the following result.

Theorem 2.4.20

Let

- Ω *be one of the complete fans $\Omega_{\mathbb{P}^2}$ or $\Omega_{\mathbb{P}^1 \times \mathbb{P}^1}$ in $\mathbb{R}^2$, and let $\mathbf{X} := \mathbf{X}(\Omega)$ denote the corresponding toric variety (i.e. $\mathbf{X} = \mathbb{P}^2$ or $\mathbf{X} = \mathbb{P}^1 \times \mathbb{P}^1$),*

- $C_1, \ldots, C_n$ *be Ω-directional tropical cycles, and let $\gamma_1, \ldots, \gamma_n \in A^*(\mathbf{X})$ be the associated cohomology classes of $\mathbf{X}$,*

- Δ *be a labelled degree with primitive direction vectors whose unlabelled degree $\delta(\Delta)$ is Ω-directional (in particular, Δ is strongly unimodular), and let $\beta \in A^{r-1}(\mathbf{X})$ be the corresponding cohomology class,*

- $a_1, \ldots, a_n$ *be non-negative integers such that $a_k = 0$ if $\dim(C_k) > 0$.*

*Then the tropical and algebro-geometric gravitational descendants are equal,
i.e.*

$$\frac{1}{\Delta!} \langle \tau_{a_1}(C_1) \cdots \tau_{a_n}(C_n) \rangle_{\Delta}^{\mathbb{R}^2} = \langle \tau_{a_1}(\gamma_1) \cdots \tau_{a_n}(\gamma_n) \rangle_{\beta}^{\mathbf{X}}.$$

Proof. First we choose a basis $B_0, \ldots, B_m$ of $Z_*(\Omega)$. This also determines a basis $\eta_0, \ldots, \eta_m$ of $A^*(\mathbf{X})$ (cf. theorem 1.1.15), and we know from the comparison to the fan displacement rule (cf. theorem 1.5.17) that

$$\deg(B_e \cdot B_f) = \deg(\eta_e \cdot \eta_f)$$

holds. This implies that, if we use WDVV equations or topological recursion with respect to these bases, then the diagonal coefficients β_{ef} appearing in the tropical and in the algebro-geometric setting coincide. Thus, using the results of the previous sections we know that the numbers $\frac{1}{\Delta!} \langle \tau_{a_1}(C_1) \cdots \tau_{a_n}(C_n) \rangle_{\Delta} = \langle \tau_{a_1}(C_1) \cdots \tau_{a_n}(C_n) \rangle_{\delta(\Delta)}$ and $\langle \tau_{a_1}(\gamma_1) \cdots \tau_{a_n}(\gamma_n) \rangle_{\beta}^{\mathbf{X}}$ satisfy a certain set of identical equations, namely the WDVV and topological recursion equations (where on the tropical side we have to be slightly more careful about i, j, k, l satisfying condition iii) and iv) of theorem 2.4.15) as well as the string and divisor equation. Therefore we can finish the proof by showing that the numbers can be computed recursively, using these equations, from some initial numbers and proving that these initial numbers coincide.

We separate the labels of the marked leaves into the sets $L \cup M \cup N = [n]$ according to the (co-)dimension of C_k as in subsection 2.4.3. First we use topological recursion to reduce the number of Psi-conditions: We pick a marked leaf x_i with $a_i > 0$ (and therefore $i \in N$) and an arbitrary pair of marked leaves x_k, x_l satisfying condition iii) of 2.4.15. If such x_k, x_l do not exist, we can add them using the divisor equation backwards with appropriate rational functions h_k, h_l. Namely, if $\mathbf{X} = \mathbb{P}^1 \times \mathbb{P}^1$ we can use $h_k = h_l = \max\{0, x, y, x + y\}$, otherwise we can use $h_k = h_l = \max\{0, x, y\}$. Note also that this choice ensures that $h_k \cdot \Delta = h_l \cdot \Delta$ is non-zero for every possible degree, so we do not divide by zero.

After eliminating all Psi-conditions in this way, we can assume $a_k = 0$ for all $k \in [n]$, i.e. we are back in the case of usual (primary) Gromov-Witten invariants. After applying the string and divisor equation we can assume that $L = M = \emptyset$ and it remains to compute invariants of the form $\langle \prod_{k=1}^{n} \tau_0(P_k) \rangle_\Delta$ for points $P_1, \ldots, P_n \in \mathbb{R}^2$. Comparing dimension shows $\#\Delta = n + 1$. Let us first consider the general case $n \geq 3$. Here we consider the one-dimensional family $F = (\tau_0(C_i)\tau_0(C_j) \prod_{k=1}^{n-1} \tau_0(P_k))_\Delta$ with arbitrary Ω-directional curves C_i, C_j such that $C_i \cdot C_j$ is non-zero and such that condition iii) of 2.4.15 is satisfied (e.g. we can choose the divisors of the functions chosen above). We let x_i, x_j be the first two marked leaves as indicated, and choose $k, l \in [n - 1]$ arbitrarily. In the corresponding WDVV equation only one extremal partition $I|J$ with $\Delta_I = 0, \Delta_J = \Delta$ does not vanish. This follows from lemma 2.4.18 and the fact that the three sums $\mathrm{codim}(P_k) + \mathrm{codim}(P_l), \mathrm{codim}(C_i) + \mathrm{codim}(P_k), \mathrm{codim}(C_j) + \mathrm{codim}(P_l)$ are greater than 2. Moreover, the only remaining extremal partition $I = \{i, j\}, J = \Delta \cup [n - 1]$ provides the term

$$\langle \tau_0(C_i)\tau_0(C_j)\tau_0(\mathbb{R}^2) \rangle_0 \cdot \langle \tau_0(P) \prod_{k=1}^{n-1} \tau_0(P_k) \rangle_\Delta = \deg(C_i \cdot C_j) \cdot \langle \prod_{k=1}^{n} \tau_0(P_k) \rangle_\Delta.$$

Hence, we can reduce the computation of $\langle \prod_{k=1}^{n} \tau_0(P_k) \rangle_\Delta$ to invariants of smaller degree. We can repeat this until we arrive at the initial invariants with $n = 1$ or $n = 2$. In these cases $\#\Delta = 2$ or $\#\Delta = 3$ and therefore the only possible degrees (up to identification via linear isomorphisms of $\mathbb{Z}^r$) are $\Delta = \{-e_1, e_1\}$ and $\Delta = \{-e_1, -e_2, e_1 + e_2\}$. In both cases, it is easy to show by direct computation that $\langle \tau_0(P_1) \rangle_\Delta = 1$ and $\langle \tau_0(P_1)\tau_0(P_2) \rangle_\Delta = 1$ hold. But now, as discussed above, the same recursion for the classical numbers proves the claim. $\qquad\square$

Remark 2.4.21 (Multiplicities of tropical curves)
The above theorem reduces the computation of the classical gravitational

descendants to the count of certain tropical curves C with multiplicities $\text{mult}(C)$ (cf. remark 2.2.14). In the above case of plane curves, an easy formula for this multiplicity exists (cf. [MR08, lemma 9.3]). Namely, if we assume general position, the multiplicity of a curve in the count is obtained as the product

$$\text{mult}(C) = \prod_V \text{mult}(V),$$

where the product runs through all vertices to which no marked leaf is adjacent and $\text{mult}(V)$ of these necessarily 3-valent vertices is the well-known vertex multiplicity introduced by Mikhalkin (cf. [Mi03, definition 2.16]). This is correct for labelled curves C, but we can as well count unlabelled curves $\tilde{C}$ (as the incidence and valence conditions do not depend on the labelling). If an unlabelled curve $\tilde{C}$ occuring in the count is automorphism-free, then there are precisely $\Delta!$ labellings C of $\tilde{C}$, and as we also divide the "labelled" invariant by $\Delta!$ to get the corresponding "unlabelled" one, the correct multiplicity of $\tilde{C}$ is $\text{mult}(\tilde{C}) = \text{mult}(C)$. In the general case, the multiplicity is given by

$$\text{mult}(\tilde{C}) = \frac{1}{\#\text{Aut}(\tilde{C})} \text{mult}(C),$$

where $\#\text{Aut}(\tilde{C})$ denotes the number of automorphisms of $\tilde{C}$.

Moreover, as well as for the usual Gromov-Witten invariants considered in [Mi03], there exists a so-called lattice path algorithm to compute these counts easily (cf. [MR08, section 9]).

Remark 2.4.22 (Rational Hurwitz numbers)
Similarly we can deal with the case $r = 1$, i.e. we can prove

$$\frac{1}{d!^2} \langle \tau_0(\mathbb{R}^1)^l \prod_{k=1}^n \tau_{a_k}(P_k) \rangle_d^{\mathbb{R}^1} = \langle \tau_0([\mathbb{P}^1])^l \prod_{k=1}^n \tau_{a_k}([pt]) \rangle_d^{\mathbb{P}^1},$$

where the left hand side is a tropical, the right hand side a classical invariant, $[pt]$ denotes the class of a point $pt \in \mathbb{P}^1$ and l, n, a_k, d denote non-negative integers. In fact, after applying the string equation, we are left with the case where $l = 0$. Now we use lemma 2.4.10 and topological recursion to reduce the number of Psi-conditions (where, if $n < 3$, we first add more marked leaves using the divisor equation). Finally, when $a_k = 0$ for all $k \in [n]$, it follows that $d = 1$ and we can compute directly $\langle \tau_0(P) \rangle_1^{\mathbb{R}^1} = 1$. For the case of the rational Hurwitz numbers $H_d^0 := \langle \tau_1([pt])^{2d-2} \rangle_d^{\mathbb{P}^1}$, this result was known before (cf. [CJM08, lemma 9.7]), but the proof is given in a different framework. In [CJM08] the result is a specialization of considerations for higher genus, not for higher dimension r as it is the case here.

Remark 2.4.23

The discussion in remark 2.4.16 and the factor $n + \#\Delta - 2$ appearing in the tropical dilaton equation 2.2.17, instead of $n - 2$ in the algebro-geometric version, show that for degrees Δ which are not strongly unimodular (if $r = 2$) and for Psi-conditions at marked leaves x_k with $\dim(C_k) > 0$, the corresponding tropical and classical invariants are in general different. For example, if we add a marked leaf that has to satisfy only a Psi-condition, the different factors in the dilaton equations immediately lead to different invariants.

Remark 2.4.24

As a final remark, let us emphasize again the strength of the theory developed in this chapter: In spite of theorem 2.4.20, it also works in higher dimensions. For example, by remark 2.3.17 and theorem 2.4.17, the same approach can be used to show that tropical and classical Gromov-Witten invariants (without Psi-classes) of $\mathbb{P}^r$, r arbitrary, coincide.

Bibliography

[Ab06]   Mohammed Abouzaid, *Morse Homology, Tropical Geometry, and Homological Mirror Symmetry for Toric Varieties*, preprint arxiv:math/0610004.

[Al09]   Lars Allermann, *Tropical intersection products on smooth varieties*, preprint arxiv:0904.2693.

[AR07]   Lars Allermann, Johannes Rau, *First steps in tropical intersection theory*, Mathematische Zeitschrift 264, 633–670 (2010); also at arxiv:0709.3705.

[AR08]   Lars Allermann, Johannes Rau, *Tropical rational equivalence on $\mathbb{R}^r$*, preprint arxiv:0811.2860.

[Be75]   D. N. Bernshtein, *The number of roots of a sytem of equations*, Funct. Anal. Appl. 9, 183–185 (1975); translated from Funktsional'nyi Analiz i Ego Prilozheniya 9, No. 3, 1–4 (1975).

[BG84]   Robert Bieri and J. R. J. Groves, *The geometry of the set of characters induced by valuations*, J. Reine Angew. Math. 347, 168–195 (1984).

[Bri89]   Michel Brion, *Groupe de Picard et nombres charactéristiques des varietés sphériques*, Duke Math. J. 58, no. 2, 397–424 (1989).

[CJM08]   Renzo Cavalieri, Paul Johnson, Hannah Markwig, *Tropical Hurwitz Numbers*, preprint arxiv:0804.0579.

[EKL04]   Manfred Einsiedler, Mikhail Kapranov, Douglas Lind, *Non-archimedean amoebas and tropical varieties*, J. Reine Angew. Math. 601, 139–157 (2006); also at arxiv:math/0408311.

[Fra08]   Marina Franz, *The tropical Kontsevich formula for toric surfaces*, diploma thesis, 2008.

[Fu84]   William Fulton, *Intersection theory*, Ergebnisse der Mathematik und ihrer Grenzgebiete, 3. Folge, Bd. 2, Springer-Verlag Berlin, XI, 470 p. (1984).

[Fu93]   William Fulton, *Introduction to toric varieties*, Annals of mathematical studies 131, Princeton, New Jersey, 157 p. (1993).

[FM08]   Marina Franz, Hannah Markwig, *Tropical enumerative invariants of $\mathbb{F}_0$ and $\mathbb{F}_2$*, preprint arxiv:0808.3452.

[FMSS95]   William Fulton, Robert MacPherson, Frank Sottile, Bernd Sturmfels, *Intersection theory on spherical varieties*, J. Algebr. Geom. 4, No. 1, 181–193 (1995).

[FP95]   William Fulton, Rahul Pandharipande, *Notes on stable maps and quantum cohomology*, Proc. Symp. Pure Math. 62, part 2, 45–96 (1997); also at arxiv:alg-geom/9608011.

[FS94]   William Fulton, Bernd Sturmfels, *Intersection Theory on Toric Varieties*, Topology 36, No. 2, 335–353 (1997); also at arxiv:alg-geom/9403002.

[Gr09]   Mark Gross, *Mirror symmetry for $\mathbb{P}^2$ and tropical geometry*, preprint arxiv:0903.1378.

[Gu06]   Walter Gubler, *Tropical varieties for non-archimedean an-alytic spaces*, Invent. Math. 169, 321–376 (2007); also at arxiv:math/0609383.

[GKM07]   Andreas Gathmann, Michael Kerber, Hannah Markwig, *Tropi-cal fans and the moduli spaces of tropical curves*, Com-pos. Math. 145, No. 1, 173–195 (2009); also at arxiv:0708.2268.

[GKZ94]   Israel Gelfand, Mikhail Kapranov, Andrei Zelevinsky, *Discrim-inants, resultants and multidimensional determinants*, Math-ematics: Theory & applications, Birkhäuser Boston, 523 p. (1994).

[GM05]   Andreas Gathmann, Hannah Markwig, *The numbers of tropical plane curves through points in general position*, J. Reine Angew. Math. 602, 155–177 (2007); also at arxiv:math/0504390.

[GM05]   Andreas Gathmann, Hannah Markwig, *Kontsevich's formula and the WDVV equations in tropical geometry*, Adv. Math. 217, No. 2, 537–560 (2008); also at arxiv:math/0509628.

[GM07]   Angela Gibney, Diane Maclagan, *Equations for Chow and Hilbert Quotients*, preprint arxiv:0707.1801.

[GZ]   Andreas Gathmann, Eva-Maria Zimmermann, *The WDVV equations in tropical geometry*, in preparation.

[H07]   Matthias Herold, *Intersection theory of the tropical moduli spa-ces of curves*, diploma thesis (2007).

[HK08]   David Helm, Eric Katz, *Monodromy filtrations and the topology of tropical varieties*, preprint arxiv:0804.3651.

[IKS04]   I. Itenberg, V. Kharlamov, and E. Shustin, *A tropical calculation of the Welschinger invariants of real toric Del Pezzo surfaces*, J. Algebraic Geom. 15, no. 2, 285–322 (2006); also at arxiv:math/0406099.

[Jo08]   Michael Joswig, *Tropical Convex Hull Computations*, Proceedings of the International Conference on Tropical and Idempotent Mathematics (to appear); also at arxiv:0809.4694.

[Ka06]   Eric Katz, *A Tropical Toolkit*, Expo. Math. 27, No. 1, 1–36 (2009); also at arxiv:math/0610878.

[Ka09]   Eric Katz, *Tropical Intersection Theory from Toric Varieties*, preprint arxiv:0907.2488.

[Kap93]   Mikhail Kapranov, *Chow quotients of Grassmannians I*, I. M. Gelfand Seminar 16, Adv. Soviet Math., 29–110 (1993); also at arxiv:alg-geom/9210002.

[Ko]   Joachim Kock, *Notes on Psi classes*, available at http://mat.uab.es/~kock/GW/notes/psi-notes.pdf.

[KM06]   Michael Kerber, Hannah Markwig, *Counting tropical elliptic plane curves with fixed j-invariant*, Comment. Math. Helv. 84, No. 2, 387–427 (2009); also at arxiv:math/0608472.

[KM07]   Michael Kerber, Hannah Markwig, *Intersecting Psi-classes on tropical $M_{0,n}$*, Int. Math. Res. Not. 2009, No. 2, 221–240 (2009); also at arxiv:0709.3953.

[KM94]   M. Kontsevich and Y. Manin, *Gromov-Witten classes, quantum cohomology and enumerative geometry*, Comm. Math. Phys. 164, 525–562 (1994).

[KV07]   Joachim Kock, Israel Vainsencher, *An Invitation to Quantum Cohomology*, Progress in Mathematics 249, Birkhäuser Boston, 159 p. (2007).

[Mi03]   Grigory Mikhalkin, *Enumerative tropical geometry in* $\mathbb{R}^2$, J. Amer. Math. Soc. 18, 313–377 (2005); also at arxiv:math/0312530.

[Mi06]   Grigory Mikhalkin, *Tropical geometry and its applications*, Int. Congress Math. Vol. II, 827–852 (2006); also at arxiv:math/0601041.

[Mi07]   Grigory Mikhalkin, *Moduli spaces of rational tropical curves*, Proceedings of the 13th Gökova geometry-topology conference, Cambridge, MA, International Press, 39–51 (2007); also at arxiv:0704.0839.

[MR08]   Hannah Markwig, Johannes Rau, *Tropical descendant Gromov-Witten invariants*, manuscripta mathematica DOI:10.1007/s00229-009-0256-5; also at arxiv:0809.1102.

[NS04]   Takeo Nishinou, Bernd Siebert, *Toric degenerations of toric varieties and tropical curves*, Duke Math. J. 135, No. 1, 1–51 (2006); also at arxiv:math/0409060.

[PS03]   Lior Pachter, Bernd Sturmfels, *Tropical Geometry of Statistical Models*, Proceedings of the National Academy of Sciences, USA, 101, 16132–7 (2004); also at arxiv:q-bio/0311009.

[R08]   Johannes Rau, *Intersections on tropical moduli spaces*, preprint arxiv:0812.3678.

[RGST]   J. Richter-Gebert, B. Sturmfels, T. Theobald, *First steps in tropical geometry*, Idempotent Mathematics and Mathematical

Physics (G.L. Litvinov and V.P. Maslov, eds.), Proceedings Vienna 2003, American Mathematical Society, Contemp. Math., (377), 2005; also at arxiv:math/0306366.

[Sm96] A. L. Smirnov, *Torus schemes over a discrete valuation ring*, St. Petersbg. Math. J. 8, No. 4, 651–659 (1997); translated from Algebra Anal. 8, No. 4, 161–172 (1996).

[Sp02] David Speyer, *Tropical Geometry*, PhD thesis, Harvard University (2002); available at http://www-math.mit.edu/~speyer.

[SS04] David Speyer, Bernd Sturmfels, *The tropical Grassmannian*, Advances in Geometry 4, No. 3, 389–411 (2004); also at arxiv:math/0304218.

[Te04] Jenia Tevelev, *Compactifications of subvarieties of tori*, Amer. J. Math. 129, no. 4 (2007); also at arxiv:math/0412329.

[Zi94] Günter M. Ziegler, *Lectures on Polytopes*, Graduate texts in mathematics 152, Springer-Verlag New York, 370 p. (1994).

[Zim07] Eva-Maria Zimmermann, *Generalizations of the tropical Kontsevich formula to higher dimensions*, diploma thesis (2007).

www.ingramcontent.com/pod-product-compliance
Lightning Source LLC
Chambersburg PA
CBHW021047210326
41598CB00016B/1128